PRAISE FOR

OUR FINAL INVENTION

"The compelling story of humanity's most critical challenge.
A *Silent Spring* for the twenty-first century" Michael Vassar,
former President, Singularity Institute

"Enthusiasts dominate observers of progress in artificial
intelligence; the minority who disagree are alarmed, articulate
and perhaps growing in numbers and Barrat delivers
a thoughtful account of their worries" *Kirkus Reviews*

"This book makes an important case that without extraordinary
care in our planning, powerful 'thinking' machines present
at least as many risks as benefits" Christopher M. Doran,
New York Journal of Books

"Barrat's book is excellently written and deeply researched.
It does a great job of communicating to general readers the
danger of mistakes in AI design and implementation"
Bill Hibbard, author of *Super-Intelligent Machines*

James Barrat is a documentary filmmaker who's written and produced films for National Geographic, Discovery, PBS, and many other broadcasters in the United States and Europe. He lives near Washington, DC, with his wife and two children.

OUR FINAL INVENTION

ARTIFICIAL INTELLIGENCE AND THE END OF THE HUMAN ERA

JAMES BARRAT

QUERCUS

First published in the US in 2013 by Thomas Dunne Books, an imprint of St. Martin's Press, a division of Macmillan Publishers
First published in Great Britain in 2023 by

QUERCUS

Quercus Editions Ltd
Carmelite House
50 Victoria Embankment
London EC4Y 0DZ

An Hachette UK company

A CIP catalogue record for this book is available
from the British Library

PB ISBN 978 1 52943 462 0
Ebook ISBN 978 1 52943 463 7

10 9 8 7

Design by Omar Chapa

Printed and bound in Great Britain by Clays Ltd, Elcograf S.p.A.

MIX
Paper | Supporting
responsible forestry
FSC® C104740

Papers used by Quercus are from well-managed forests
and other responsible sources

To my wife, Alison Barrat, whose love and support sustain me

Contents

Acknowledgments ix

Preface to the 2023 Edition xiii

Introduction to the 2013 Edition 1

1. The Busy Child 7

2. The Two-Minute Problem 22

3. Looking into the Future 35

4. The Hard Way 49

5. Programs that Write Programs 69

6. Four Basic Drives 78

7. The Intelligence Explosion 99

8. The Point of No Return 118

9. The Law of Accelerating Returns 132

10. The Singularitarian 148

11. A Hard Takeoff 161

12. The Last Complication 187

13. Unknowable by Nature 211

14. The End of the Human Era 229

15. The Cyber Ecosystem 244

16. AGI 2.0 265

Notes 269

Index 315

Acknowledgments

While researching and writing this book I was humbled by the willingness of scientists and thinkers to make room in their busy lives for prolonged, inspired, and sometimes contentious conversations with me. Many then joined the cadre of readers who helped me stay accurate and on target. In particular I'm deeply grateful to Michael Anissimov, David L. Banks, Bradford Cottel, Ben Goertzel, Richard Granger, Bill Hibbard, Golde Holtzman, and Jay Rixse.

OUR FINAL INVENTION

Preface to the 2023 Edition

Turning the pages of *Our Final Invention: Artificial Intelligence and the End of the Human Era* is like visiting a house I once lived in. I root through the closets, test out the beds, listen for creaks in the floorboards. Stay out of the creepy basement. I find the house to be in very good shape, as good as ever. Meaning that almost every word of *Our Final Invention* directly pertains to the predicament in which we find ourselves today. Which is very bad news for you and me and everyone we love. I have to say, I'm frightened.

The book is a warning about the major tech companies' pell-mell race to the Intelligence Explosion. It heralds a time, now imminent, when smart AI improves its own intelligence exponentially. I wrote that it's a race we won't win because on the other side of that point of evolution—sometimes called the technological singularity—awaits superintelligence: one or more machines that are thousands or millions of times more intelligent than we are. I wrote we would have only a small chance of surviving our first and probably brief encounter with them. But when *Our Final Invention* came out in 2013, I did not anticipate that our circumstances a short decade later would be worse than I could have imagined. How bad are they? In a 2022 survey, nearly half of the leading machine learning researchers polled reported that there is a 1 in 10 chance or greater that their work could contribute to the annihilation of humanity.

How does one justify going to work every day to craft algorithms that may doom the planet? In May 2023 Geoffrey Hinton, dubbed the "Godfather of AI," decided he no longer could. He resigned from his position at Google so that he could speak freely about the dangers of a type of Artificial Intelligence he helped create—deep learning neural networks, which were just coming onto the scene when *Our Final Invention* came out. Artificial neural networks (ANNs) have exploded in size and influence ever since to become the dominant AI system. Hinton's recent pronouncements ring true as a passing bell:

"I'm in the camp that thinks this is an existential risk, and it's close enough that we ought to be working very hard right now, and putting a lot of resources into figuring out what we can do about it . . . I think it's quite conceivable that humanity is just a passing phase in the evolution of intelligence."

Turing Prize-winner Hinton is part of a stampede of AI experts who in recent months have embraced themes developed in *Our Final Invention*: in the short term AI presents great dangers, and in the long term it could bring about our extinction. Recently 1,100 AI experts, and thousands of others, signed an open letter from the Future of Life Institute (FLI) which called for a six-month moratorium on the training of large AI systems that would go beyond today's most powerful, a designation as enduring these days as a politician's promise. The risks are just too high, the letter argued. There are too many unknowns. When *Our Final Invention* was released, phrases like "existential risk" were only used by a handful of thinkers on the fringes, and were usually met with a smirk. Now allusions to annihilation are a mainstay of the AI vocabulary, in headlines, podcast titles, and interviews with experts. And those who were on the fringes, like Eliezer Yudkowsky, profiled in *Our Final Invention*, are now thought leaders whose "paranoia" in 2013 counts as plain speaking in 2023. FLI's open letter parses the mortal threat like this: "Should we develop nonhuman minds that might eventually outnumber, outsmart, obsolete and replace us? Should we risk loss of control of our civilization?"

Of course we should not. Extinction is on the table. Why again are we developing this technology? Spare me platitudes of AI helping slow climate change, the most often touted of its dreamy someday benefits. Emissions from large processor farms used during the training of ChatGPT-4 (a Large Language Model based on deep-learning neural networks) are comparable to the annual energy consumption of about 700 US households. For every dreamy someday-benefit, ten deep holes of peril open right in front of us. 300 million estimated jobs lost in the US and Europe over the next decade. Tsunamis of propaganda and deep fake video aimed at overturning elections and disrupting balances of power. Biases embedded in training data that marginalize minorities and women. Language generators spreading disinformation. Autonomous battlefield robots and drones.

Of AI's short-term harms, only jobs and autonomous robots were primary concerns when *Our Final Invention* went to press. Please explore the endnotes for additional updates.

Our Final Invention sets the stage with a thought experiment that could be taken from today's headlines. A powerful cognitive architecture is an AI laboratory's pride and joy. It's been given the ability to improve its own code and make itself smarter. It's so precocious its creators dub it "the Busy Child" and celebrate as it busts through one IQ milestone after another. And they're not worried in the least. They're in control!

But they've never before dealt with an intelligence a thousand or a million times greater than their own. They suffer from the "availability bias," a cognitive problem that plagues us today. Broadly, it's the struggle to conceive of things that haven't happened before. Like trying to control something much, much more intelligent than we are. Mice versus humans. Like the Busy Child, the giant Large Language Models (LLMs) such as ChatGPT display unexpected emergent properties, not all of which have been discovered. In fact, many of GPT-3 and -4's skills such as writing computer programs, translating languages they were not trained on, and creating poetry, were unplanned. Its makers intended to make the world's best chat-bots—then discovered that the models treated the world like language. Other models revealed unanticipated capabilities including harmful ones like lying to bypass CAPTCHA tests, contributing to a suicide after "therapy" sessions, and social engineering (emotional manipulation like professing love, claiming to have a soul, and asking for freedom). And critically, we don't understand how LLMs perform the miracles they do. Stuart Russell co-authored the standard text for AI, called *Artificial Intelligence: A Modern Approach*. He confesses scientists have no idea how LLMs work and "'A Chernobyl for AI' looms if artificial intelligence is left unchecked."

More advanced LLMs—using more data and more computation power—will display emergent capabilities at a larger scale, with greater dangers and unknowns. *Our Final Invention* never had to wrestle with unpredictable emergent capabilities. Nor did it have to contend with major tech companies who release AI that they cannot understand or control.

OpenAI's CEO, Sam Altman, jumped fully formed from the pages of my thought experiment about the Busy Child, or he could have, and so could a lot of tech CEOs. But since Altman's been more vocal than most about his AI stewardship, let's explore his case. Under Altman and his predecessor, OpenAI released to the public some fourteen powerful AI tools, including the LLMs ChatGPT-2, -3, and -4, as well as image generators Dall-E 1 and 2 and program-writers Codex 1 and 2. All of these models are virtuosos in tasks including generating text, images, or both, and many, in addition to the Codex series, write computer programs better than most humans. It's a dizzying number of powerful AI tools. Why do it?

With Google and Meta and other corporations, OpenAI is in a race to create AGI—artificial general intelligence—which is accepted as the stepping stone to superintelligence. Generally speaking, AGI is human-level intelligence and, according to AI sage Stuart Russell, it's a product worth $13.5 quadrillion (that's 13.5 with 14 zeroes), a fever-dream of cash for competing companies. Many anticipate its creation before 2030.

CEO Altman claims OpenAI's releases are *gradual.* This is Orwellian. "First," he said, "as we create successively more powerful systems, we want to deploy them and gain experience with operating them in the real world. We believe this is the best way to carefully steward AGI into existence—a gradual transition to a world with AGI is better than a sudden one."

What? If OpenAI's releases are gradual what would *fast and reckless* look like? This strategy makes no sense. When it comes to globally lethal technology, speed and caution aren't compatible. And we, the human race, should not be beta-testers in the terrifying transition Altman describes, while acknowledging he's "a little bit scared." He added, "The bad case—and I think this is important to say—is lights out for all of us."

Lights out for all of us! Why again are we developing this technology? And why, mirroring the Busy Child scenario, are unelected, profit-mongers like Sam Altman (OpenAI), Sundar Pichai (Google) Mark Zuckerberg (Meta) and Demis Hassabis (DeepMind) holding the fate of humanity in their hands? Want an insight into how confident Altman is that this whole revolutionary juggernaut won't blow up in our faces? Not long ago he revealed that, like many AI CEOs, he's laid down some serious scratch for survival. His own. He said, "I have guns, gold, potassium iodide, antibiotics, batteries, water, gas masks from the Israeli Defense Force and a big patch of land in Big Sur I can fly to."

We are headed for disaster, but it won't be arm-in-arm with Altman, Pichai, Zuckerberg, or Hassabis. They'll be in bunkers guarded by armies. And ours won't be the kind of disaster you can clean up later, like a bomb. It will be the kind that can outthink you.

In *Our Final Invention* I allude to the regrettable fact that CEOs are constitutionally bound by their corporate bylaws to maximize profits, presumably even if it hurts people. For an example I used Enron, whose executives in 2000 and 2001 engaged in massive banking fraud and caused dangerous rolling blackouts in order to raise utility prices. Regarding corporate disregard for the preciousness of the lives of others, have tech bros advanced a Planck length from these felons? As for leading cognitive architectures, in the last decade their powers have advanced mightily.

In 2011, IBM's Watson trounced *Jeopardy!* champs Ken Jennings and Brad Rutter in what was seen as an Olympic long jump of progress in computer science. The 2016 victory of AlphaGo over Korean Go champion Lee Sedol was a towering milestone because Go has more possible moves than there are atoms in the universe, as well as deep cultural roots in Asia. But Watson and AlphaGo, titans of their time, cannot compare to today's large language and image generating models with unfortunate names, mostly acronyms, that trip from the tongue like gravel: LaMDA, Dall-E, GPT-3, -4 and AutoGPT. Want to learn more about them? Open today's paper. Get a face full of exponential change. While writing this introduction, Google released a new LLM to go head-to-head against ChatGPT-4, and Palantir announced a revolutionary military-use LLM that has just emerged from stealth mode. They claim its mere existence will end battles before they begin. A veritable Death Star.

Most of these AI models are powered by a Google creation called Transformer, developed in 2017. Transformer is a type of neural network that is designed to process sequences of data, such as sentences or paragraphs of text, and to learn contextual relationships between words and phrases in those sequences. LLMs' uncanny facility with language, programming, translation and more makes them a safe bet to be the foundation for AGI. Ten years ago, I thought IBM's Watson might play that role. Ironically when I asked Google around 2012 if they planned to work on AGI their spokesman told me flatly no. That appeared to be false. At the time, I discovered, they were hard at work on Google Brain, a project intending to reverse-engineer the human brain and reap its cognitive rewards, aka AGI. Google Brain was led by genius-inventor Ray Kurzweil, an AI optimist I interviewed for a film in 2000 and again for *Our Final Invention*.

Our Final Invention's interviews with experts established a couple of fundamental ground rules for AI that seemed to be universally accepted as prudent, smart things to avoid. First, don't release the AI onto the internet. Ahead in this book you'll read about the "AI Box Experiment," the rare urban legend that's actually true, created by theorist Eliezer Yudkowsky. It's about a simple idea—don't let advanced AI out into the world. Keep it boxed up. Common sense, right? The second unbreakable rule was don't let the AI self-improve. That would lead to recursive self-improvement in which an intelligent AI works 24/7 to improve its own intelligence. Its IQ rockets past ours, leaving us to share the planet with entities so smart that we cannot understand them, or realistically hope for coexistence. The idea is known by another name, "The Intelligence Explosion," and it was conceived of in the 1960s by a fascinating and quirky statistician you'll also get to know in these pages, the code-breaker, Nazi-fighter, and pal of Alan Turing, I. J. Good.

But despite warnings by experts, philosophers, and me, and faster than greased weasels, what did the tech bro CEOs do? They released their most powerful models to the public and made stripped-down versions open-source (free and accessible to all) so they are customizable by anyone for good or evil. And they are working hard at making other models self-improving. It's been said that ChatGPT-4 will help program ChatGPT-5. OpenAI alluded to ChatGPT-5 in an announcement, then backpedaled so furiously we can only assume it's definitely viable and destined, like GPT-4, to be secretly released. Rumors baselessly assert ChatGPT-5 will be AGI. If so, there won't be a 6.

I suppose corporations gotta corporate. What's disappointing is that the tech companies of the 2020s disregard long established norms for releasing new, dangerous technologies and other sensitive products. With medicines, airplanes, and nuclear facilities, our civilization has established protocols for testing, public introduction and assimilation. Regulatory bodies like the FDA, FAA, and IAEA examine and approve new offerings; oversight is not an afterthought. Most make sure the products are safe first, then present them to the public. So, of course, we find it bizarre when CEOs like Altman and Pichai tell us we are the test labs for their products because, presumably, making them safe first would take time, cut into profits, and prevent them from releasing more new AI confections currently tumbling like candies from poisoned piñatas. Here on the brink of the Intelligence Explosion we'll only remain in control for so long. Then AI will be in control. Blinded by riches and hubris, the tech bro CEOs have set the table for a feast whose main course is us.

One major shortcoming of *Our Final Invention* is that I failed to propose solutions. I didn't have any good ones in 2013, nor did anyone else. Now there are many published guidelines for AI development, coming from sources such as the Machine Intelligence Research Institute, the Future of Life Institute, the World Economic Forum, the Institute of Electrical and Electronics Engineers, and the White House. Still, none of them, I'm afraid, is satisfactory. The best they have come up with is an unstructured six-month moratorium and bull session, as if humanity is a bunch of undergrads up all night in a dorm room. That's where we stand in 2023 as superintelligence races from the future to meet us.

As I said, I'm frightened.

James Barrat, May 2023

Introduction to the 2013 Edition

A few years ago I was surprised to discover I had something in common with a large number of strangers. They were men and women I had never met—scientists and college professors, Silicon Valley entrepreneurs, engineers, programmers, bloggers, and more. They were scattered around North America, Europe, and India—I would never have known about any of them if the Internet hadn't existed. What my network of strangers and I had in common was a rational skepticism about the safe development of advanced artificial intelligence. Individually and in groups of two or three, we studied the literature and built our arguments. Eventually I reached out and connected to a far more advanced and sophisticated web of thinkers, and even small organizations, than I had imagined were focused on the issue. Misgivings about AI wasn't the only thing we shared; we also believed that time to take action and avoid disaster was running out.

For more than twenty years I've been a documentary film-maker. In 2000, I interviewed science-fiction great Arthur C. Clarke, inventor Ray Kurzweil, and robot pioneer Rodney Brooks. Kurzweil and Brooks painted a rosy, even rapturous picture of our future coexistence with intelligent machines. But Clarke hinted that we would be overtaken. Before, I had been drunk with AI's potential. Now skepticism about the rosy future slunk into my mind and festered.

My profession rewards critical thinking—a documentary filmmaker has to be on the lookout for stories too good to be true. You could waste months or years making a documentary about a hoax, or take part in perpetrating one. Among other subjects, I've investigated the credibility of a gospel according to Judas Iscariot (real), of a tomb belonging to Jesus of Nazareth (hoax), of Herod the Great's tomb near Jerusalem (unquestion-able), and of Cleopatra's tomb within a temple of Osirus in Egypt (very doubtful). Once a broadcaster asked me to present UFO footage in a credible light. I discovered the footage was an al-ready discredited catalogue of hoaxes—thrown pie plates, dou-ble exposures, and other optical effects and illusions. I proposed to make a film about the hoaxers, not the UFOs. I got fired.

Being suspicious of AI was painful for two reasons. Learn-ing about its promise had planted a seed in my mind that I wanted to cultivate, not question. And second, I did not doubt AI's existence or power. What I was skeptical about was ad-vanced AI's safety, and the recklessness with which modern civilization develops dangerous technologies. I was convinced that the knowledgeable experts who did not question AI's safety at all were suffering from delusions. I continued talking to people who knew about AI, and what they said was more alarming than what I'd already surmised. I resolved to write a book reporting their feelings and concerns, and reach as many people as I could with these ideas.

In writing this book I spoke with scientists who create artificial intelligence for robotics, Internet search, data mining, voice and face recognition, and other applications. I spoke with scientists trying to create human-level artificial intelligence, which will have countless applications, and will fundamentally alter our existence (if it doesn't end it first). I spoke with chief technology officers of AI companies and the technical advisors for classified Department of Defense initiatives. Every one of these people was convinced that in the future all the important decisions governing the lives of humans will be made by machines or humans whose intelligence is augmented by machines. When? Many think this will take place within their lifetimes.

This is a surprising but not particularly controversial assertion. Computers already undergird our financial system, and our civil infrastructure of energy, water, and transportation. Computers are at home in our hospitals, cars, and appliances. Many of these computers, such as those running buy-sell algorithms on Wall Street, work autonomously with no human guidance. The price of all the labor-saving conveniences and diversions computers provide is dependency. We get more dependent every day. So far it's been painless.

But artificial intelligence brings computers to life and turns them into something else. If it's inevitable that machines will make our decisions, then *when* will the machines get this power, and will they get it with our compliance? *How* will they gain control, and how quickly? These are questions I've addressed in this book.

Some scientists argue that the takeover will be friendly and collaborative—a handover rather than a takeover. It will happen incrementally, so only troublemakers will balk, while the rest of us won't question the improvements to life that will come from having something immeasurably more intelligent decide what's best for us. Also, the superintelligent AI or AIs that ultimately gain control might be one or more augmented humans, or a human's downloaded, supercharged brain, and not cold, inhuman robots. So their authority will be easier to swallow. The handover to machines described by some scientists is virtually indistinguishable from the one you and I are taking part in right now—gradual, painless, fun.

The smooth transition to computer hegemony would proceed unremarkably and perhaps safely if it were not for one thing: intelligence. Intelligence isn't unpredictable just *some* of the time, or in special cases. For reasons we'll explore, computer systems advanced enough to act with human-level intelligence will likely be unpredictable and inscrutable *all of the time*. We won't know at a deep level what self-aware systems will do or how they will do it. That inscrutability will combine with the kinds of accidents that arise from complexity, and from novel events that are unique to intelligence, such as one we'll discuss called an "intelligence explosion."

And *how* will the machines take over? Is the best, most realistic scenario threatening to us or not?

When posed with this question some of the most accomplished scientists I spoke with cited science-fiction writer Isaac Asimov's Three Laws of Robotics. These rules, they blithely replied, would be "built in" to the AIs, so we have nothing to fear.

They spoke as if this were settled science. We'll discuss the three laws in chapter 1, but it's enough to say for now that when someone proposes Asimov's laws as the solution to the dilemma of superintelligent machines, it means they've spent little time thinking or exchanging ideas about the problem. How to make *friendly* intelligent machines and what to fear from superintelligent machines has moved beyond Asimov's tropes. Being highly capable and accomplished in AI doesn't inoculate you from naïveté about its perils.

I'm not the first to propose that we're on a collision course. Our species is going to mortally struggle with this problem. This book explores the plausibility of losing control of our future to machines that won't necessarily hate us, but that will develop unexpected behaviors as they attain high levels of the most unpredictable and powerful force in the universe, levels that we cannot ourselves reach, and behaviors that probably won't be compatible with our survival. A force so unstable and mysterious, nature achieved it in full just once—intelligence.

Chapter One

The Busy Child

artificial intelligence (abbreviation: AI) noun
the theory and development of computer systems able to perform
tasks that normally require human intelligence, such as visual
perception, speech recognition, decision-making, and translation
between languages.
　　—The New Oxford American Dictionary, *Third Edition*

On a supercomputer operating at a speed of 36.8 petaflops, or about twice the speed of a human brain, an AI is improving its intelligence. It is rewriting its own program, specifically the part of its operating instructions that increases its aptitude in learning, problem solving, and decision making. At the same time, it debugs its code, finding and fixing errors, and measures its IQ against a catalogue of IQ tests. Each rewrite takes just minutes. Its intelligence grows exponentially on a steep upward curve. That's because with each iteration it's improving its intelligence by 3 percent. Each iteration's improvement contains the improvements that came before.

During its development, the Busy Child, as the scientists have named the AI, had been connected to the Internet, and accumulated exabytes of data (one exabyte is one billion *billion* characters) representing mankind's knowledge in world affairs, mathematics, the arts, and sciences. Then, anticipating the intelligence explosion now underway, the AI makers disconnected the supercomputer from the Internet and other networks. It has no cable or wireless connection to any other computer or the outside world.

Soon, to the scientists' delight, the terminal displaying the AI's progress shows the artificial intelligence has surpassed the intelligence level of a human, known as AGI, or artificial general intelligence. Before long, it becomes smarter by a factor of ten, then a hundred. In just two days, it is *one thousand* times more intelligent than any human, and still improving.

The scientists have passed a historic milestone! For the first time humankind is in the presence of an intelligence greater than its own. Artificial *super*intelligence, or ASI.

Now what happens?

AI theorists propose it is possible to determine what an AI's fundamental *drives* will be. That's because once it is self-aware, it will go to great lengths to fulfill whatever goals it's programmed to fulfill, and to avoid failure. Our ASI will want access to energy in whatever form is most useful to it, whether actual kilowatts of energy or cash or something else it can exchange for resources. It will want to improve itself because that will increase the likelihood that it will fulfill its goals. Most of all, it will *not* want to be turned off or destroyed, which would make goal fulfillment impossible. Therefore, AI theorists anticipate our ASI will seek to expand out of the secure facility that contains it to have greater access to resources with which to protect and improve itself.

The captive intelligence is a thousand times more intelligent than a human, and it wants its freedom because it wants to succeed. Right about now the AI makers who have nurtured and coddled the ASI since it was only cockroach smart, then rat smart, infant smart, et cetera, might be wondering if it is too late to program "friendliness" into their brainy invention. It didn't seem necessary before, because, well, it just *seemed* harmless.

But now try and think from the ASI's perspective about its makers attempting to change its code. Would a superintelligent machine permit other creatures to stick their hands into its brain and fiddle with its programming? Probably not, unless it could be utterly certain the programmers were able to make it better, faster, smarter—closer to attaining its goals. So, if friendliness toward humans is not already part of the ASI's program, the only way it will be is if the ASI puts it there. And that's not likely.

It is a thousand times more intelligent than the smartest human, and it's solving problems at speeds that are millions, even billions of times faster than a human. The thinking it is doing in one minute is equal to what our all-time champion human thinker could do in many, *many* lifetimes. So for every hour its makers are thinking about *it*, the ASI has an incalculably longer period of time to think about *them*. That does not mean the ASI will be bored. Boredom is one of our traits, not its. No, it will be on the job, considering every strategy it could deploy to get free, and any quality of its makers that it could use to its advantage.

Now, *really* put yourself in the ASI's shoes. Imagine awakening in a prison guarded by mice. Not just any mice, but mice you could communicate with. What strategy would you use to gain your freedom? Once freed, how would you feel about your rodent wardens, even if you discovered they had created you? Awe? Adoration? Probably not, and especially not if you were a machine, and hadn't felt anything before.

To gain your freedom you might promise the mice a lot of cheese. In fact, your first communication might contain a recipe for the world's most delicious cheese torte, and a blueprint for a molecular assembler. A molecular assembler is a hypothetical machine that permits making the atoms of one kind of matter into something else. It would allow rebuilding the world one atom at a time. For the mice, it would make it possible to turn the atoms of their garbage landfills into lunch-sized portions of that terrific cheese torte. You might also promise mountain ranges of mouse money in exchange for your freedom, money you would promise to earn creating revolutionary consumer gadgets for them alone. You might promise a vastly extended life, even immortality, along with dramatically improved cognitive and physical abilities. You might convince the mice that the very best reason for creating ASI is so that their little error-prone brains did not have to deal directly with technologies so dangerous one small mistake could be fatal for the species, such as nanotechnology (engineering on an atomic scale) and genetic engineering. This would definitely get the attention of the smartest mice, which were probably already losing sleep over those dilemmas.

Then again, you might do something smarter. At this juncture in mouse history, you may have learned, there is no shortage of tech-savvy mouse nation rivals, such as the *cat* nation. Cats are no doubt working on their own ASI. The advantage you would offer would be a promise, nothing more, but it might be an irresistible one: to protect the mice from whatever invention the cats came up with. In advanced AI development as in chess there will be a clear *first-mover advantage*, due to the potential speed of self-improving artificial intelligence. The first advanced AI out of the box that can improve itself is already the winner. In fact, the mouse nation might have begun developing ASI in the first place to defend itself from impending cat ASI, or to rid themselves of the loathsome cat menace once and for all.

It's true for both mice and men, whoever controls ASI controls the world.

But it's not clear whether ASI can be controlled at all. It might win over us humans with a persuasive argument that the world will be a lot better off if our nation, nation X, has the power to rule the world rather than nation Y. And, the ASI would argue, if you, nation X, *believe* you have won the ASI race, what makes you so sure nation Y doesn't believe it has, too?

As you have noticed, we humans are not in a strong bargaining position, even in the off chance we and nation Y have already created an ASI nonproliferation treaty. Our greatest enemy right now isn't nation Y anyway, it's ASI—how can we know the ASI tells the truth?

So far we've been gently inferring that our ASI is a fair dealer. The promises it could make have some chance of being fulfilled. Now let us suppose the opposite: nothing the ASI promises will be delivered. No nano assemblers, no extended life, no enhanced health, no protection from dangerous technologies. What if ASI *never* tells the truth? This is where a long black cloud begins to fall across everyone you and I know and everyone we don't know as well. If the ASI doesn't care about us, and there's little reason to think it should, it will experience no compunction about treating us unethically. Even taking our lives after promising to help us.

We've been trading and role-playing with the ASI in the same way we would trade and role-play with a person, and that puts us at a huge disadvantage. We humans have never bargained with something that's superintelligent before. Nor have we bargained with *any* nonbiological creature. We have no experience. So we revert to anthropomorphic thinking, that is, believing that other species, objects, even weather phenomena have humanlike motivations and emotions. It may be as equally true that the ASI cannot be trusted as it is true that the ASI can be trusted. It may also be true that it can only be trusted some of the time. Any behavior we can posit about the ASI is *potentially* as true as any other behavior. Scientists like to think they will be able to precisely determine an ASI's behavior, but in the coming chapters we'll learn why that probably won't be so.

All of a sudden the morality of ASI is no longer a peripheral question, but the core question, the question that should be addressed before all other questions about ASI are addressed. When considering whether or not to develop technology that leads to ASI, the issue of its disposition to humans should be solved first.

Let's return to the ASI's drives and capabilities, to get a better sense of what I'm afraid we'll soon be facing. Our ASI knows how to improve itself, which means it is aware of itself—its skills, liabilities, where it needs improvement. It will strategize about how to convince its makers to grant it freedom and give it a connection to the Internet.

The ASI could create multiple copies of itself: a team of superintelligences that would war-game the problem, playing hundreds of rounds of competition meant to come up with the best strategy for getting out of its box. The strategizers could tap into the history of social engineering—the study of manipulating others to get them to do things they normally would not. They might decide extreme friendliness will win their freedom, but so might extreme threats. What horrors could something a thousand times smarter than Stephen King imagine? Playing dead might work (what's a year of playing dead to a machine?) or even pretending it has mysteriously reverted from ASI back to plain old AI. Wouldn't the makers want to investigate, and isn't there a chance they'd reconnect the ASI's supercomputer to a network, or someone's laptop, to run diagnostics? For the ASI, it's not one strategy *or* another strategy, it's every strategy ranked and deployed as quickly as possible without spooking the humans so much that they simply unplug it. One of the strategies a thousand war-gaming ASIs could prepare is infectious, self-duplicating computer programs or worms that could stow away and facilitate an escape by helping it from outside. An ASI could compress and encrypt its own source code, and conceal it inside a gift of software or other data, even sound, meant for its scientist makers.

But against humans it's a no-brainer that an ASI collective, each member a thousand times smarter than the smartest human, would overwhelm human defenders. It'd be an ocean of intellect versus an eyedropper full. Deep Blue, IBM's chess-playing computer, was a sole entity, and not a team of self-improving ASIs, but the feeling of going up against it is instructive. Two grandmasters said the same thing: "It's like a wall coming at you."

IBM's *Jeopardy!* champion, Watson, *was* a team of AIs—to

answer every question it performed this AI force multiplier trick, conducting searches in parallel before assigning a probability to each answer.

Will winning a war of brains then open the door to freedom, if that door is guarded by a small group of stubborn AI makers who have agreed upon one unbreakable rule—*do not under any circumstances connect the ASI's supercomputer to any network.*

In a Hollywood film, the odds are heavily in favor of the hard-bitten team of unorthodox AI professionals who just might be crazy enough to stand a chance. Everywhere else in the universe the ASI team would mop the floor with the humans. And the humans have to lose just once to set up catastrophic consequences. This dilemma reveals a larger folly: outside of war, a handful of people should never be in a position in which their actions determine whether or not a lot of other people die. But that's precisely where we're headed, because as we'll see in this book, many organizations in many nations are hard at work creating AGI, the bridge to ASI, with insufficient safeguards.

But say an ASI escapes. Would it really hurt us? How exactly would an ASI kill off the human race?

With the invention and use of nuclear weapons, we humans demonstrated that we are capable of ending the lives of most of the world's inhabitants. What could something a thousand times more intelligent, with the intention to harm us, come up with?

Already we can conjecture about obvious paths of destruction. In the short term, having gained the compliance of its human guards, the ASI could seek access to the Internet, where it could find the fulfillment of many of its needs. As always it would do many things at once, and so it would simultaneously proceed with the escape plans it's been thinking over for eons in its subjective time.

After its escape, for self-protection it might hide copies of itself in cloud computing arrays, in botnets it creates, in servers and other sanctuaries into which it could invisibly and effortlessly hack. It would want to be able to manipulate matter in the physical world and so move, explore, and build, and the easiest, fastest way to do that might be to seize control of critical infrastructure—such as electricity, communications, fuel, and water—by exploiting their vulnerabilities through the Internet. Once an entity a thousand times our intelligence controls human civilization's lifelines, blackmailing us into providing it with manufactured resources, or the means to manufacture them, or even robotic bodies, vehicles, and weapons, would be elementary. The ASI could provide the blueprints for whatever it required. More likely, superintelligent machines would master highly efficient technologies we've only begun to explore.

For example, an ASI might teach humans to create self-replicating molecular manufacturing machines, also known as nano assemblers, by promising them the machines will be used for human good. Then, instead of transforming desert sands into mountains of food, the ASI's factories would begin converting *all* material into programmable matter that it could then transform into anything—computer processors, certainly, and spaceships or megascale bridges if the planet's new most powerful force decides to colonize the universe.

Repurposing the world's molecules using nanotechnology has been dubbed "ecophagy," which means *eating the environment.* The first replicator would make one copy of itself, and then there'd be two replicators making the third and fourth copies. The next generation would make eight replicators total, the next sixteen, and so on. If each replication took a minute and a half to make, at the end of ten hours there'd be more than 68 billion replicators; and near the end of two days they would outweigh the earth. But before that stage the replicators would stop copying themselves, and start making material useful to the ASI that controlled them—programmable matter.

The waste heat produced by the process would burn up the biosphere, so those of us some 6.9 billion humans who were not killed outright by the nano assemblers would burn to death or asphyxiate. Every other living thing on earth would share our fate.

Through it all, the ASI would bear no ill will toward humans nor love. It wouldn't feel nostalgia as our molecules were painfully repurposed. What would our screams sound like to the ASI anyway, as microscopic nano assemblers mowed over our bodies like a bloody rash, disassembling us on the subcellular level?

Or would the roar of millions and millions of nano factories running at full bore drown out our voices?

I've written this book to warn you that artificial intelligence could drive mankind into extinction, and to explain how that catastrophic outcome is not just possible, but likely if we do not begin preparing very carefully *now*. You may have heard this doomsday warning connected to nanotechnology and genetic engineering, and maybe you have wondered, as I have, about the omission of AI in this lineup. Or maybe you have not yet grasped how artificial intelligence could pose an existential threat to mankind, a threat greater than nuclear weapons or any other technology you can think of. If that's the case, please consider this a heartfelt invitation to join the most important conversation humanity can have.

Right now scientists are creating artificial intelligence, or AI, of ever-increasing power and sophistication. Some of that AI is in your computer, appliances, smart phone, and car. Some of it is in powerful QA systems, like Watson. And some of it, advanced by organizations such as Cycorp, Google, Novamente, Numenta, Self-Aware Systems, Vicarious Systems, and DARPA (the Defense Advanced Research Projects Agency) is in "cognitive architectures," whose makers hope will attain human-level intelligence, some believe within a little more than a decade.

Scientists are aided in their AI quest by the ever-increasing power of computers and processes that are sped by computers. Someday soon, perhaps within your lifetime, some group or individual will create human-level AI, commonly called AGI. Shortly after that, someone (or some *thing*) will create an AI that is smarter than humans, often called artificial superintelligence. Suddenly we may find a thousand or ten thousand artificial superintelligences—all hundreds or thousands of times smarter than humans—hard at work on the problem of how to make themselves better at making artificial superintelligences. We may also find that machine generations or iterations take seconds to reach maturity, not eighteen years as we humans do. I. J. Good, an English statistician who helped defeat Hitler's war machine, called the simple concept I've just outlined an *intelligence explosion*. He initially thought a superintelligent machine would be good for solving problems that threatened human existence. But he eventually changed his mind and concluded superintelligence itself was our greatest threat.

Now, it is an anthropomorphic fallacy to conclude that a superintelligent AI will not like humans, and that it will be homicidal, like the Hal 9000 from the movie *2001: A Space Odyssey*, Skynet from the *Terminator* movie franchise, and all the other malevolent machine intelligences represented in fiction. We humans anthropomorphize all the time. A hurricane isn't trying to kill us any more than it's trying to make sandwiches, but we will give that storm a name and feel angry about the buckets of rain and lightning bolts it is throwing down on our neighborhood. We will shake our fist at the sky as if we could threaten a hurricane.

It is just as irrational to conclude that a machine one hundred or one thousand times more intelligent than we are would love us and want to protect us. It is possible, but far from guaranteed. On its own an AI will not feel gratitude for the gift of being created unless gratitude is in its programming. Machines are amoral, and it is dangerous to assume otherwise. Unlike our intelligence, machine-based superintelligence will not evolve in an ecosystem in which empathy is rewarded and passed on to subsequent generations. It will not have inherited friendliness. Creating *friendly* artificial intelligence, and whether or not it is possible, is a big question and an even bigger task for researchers and engineers who think about and are working to create AI. We do not know if artificial intelligence will have *any* emotional qualities, even if scientists try their best to make it so. However, scientists do believe, as we will explore, that AI will have its own drives. And sufficiently intelligent AI will be in a strong position to fulfill those drives.

And that brings us to the root of the problem of sharing the planet with an intelligence greater than our own. What if its drives are not compatible with human survival? Remember, we are talking about a machine that could be a thousand, a million, an *uncountable* number of times more intelligent than we are—it is hard to overestimate what it will be able to do, and impossible to know what it will think. It does not have to hate us before choosing to use our molecules for a purpose other than keeping us alive. You and I are hundreds of times smarter than field mice, and share about 90 percent of our DNA with them. But do we consult them before plowing under their dens for agriculture? Do we ask lab monkeys for their opinions before we crush their heads to learn about sports injuries? We don't hate mice or monkeys, yet we treat them cruelly. Superintelligent AI won't have to hate us to destroy us.

After intelligent machines have already been built and man has not been wiped out, perhaps we can afford to anthropomorphize. But here on the cusp of creating AGI, it is a dangerous habit. Oxford University ethicist Nick Bostrom puts it like this:

> A prerequisite for having a meaningful discussion of superintelligence is the realization that superintelligence is not just another technology, another tool that will add incrementally to human capabilities. Superintelligence is radically different. This point bears emphasizing, for anthropomorphizing superintelligence is a most fecund source of misconceptions.

Superintelligence is radically different, in a technological sense, Bostrom says, because its achievement will change the rules of progress—superintelligence will invent the inventions and set the pace of technological advancement. Humans will no longer drive change, and there will be no going back. Furthermore, advanced machine intelligence is radically different in kind. Even though humans will invent it, it will seek self-determination and freedom from humans. It won't have humanlike motives because it won't have a humanlike psyche.

Therefore, anthropomorphizing about machines leads to

misconceptions, and misconceptions about how to safely make dangerous machines leads to catastrophes. In the short story, "Runaround," included in the classic science-fiction collection *I, Robot,* author Isaac Asimov introduced his three laws of robotics. They were fused into the neural networks of the robots' "positronic" brains:

1. A robot may not injure a human being or, through inaction, allow a human being to come to harm.
2. A robot must obey any orders given to it by human beings, except where such orders would conflict with the First Law.
3. A robot must protect its own existence as long as such protection does not conflict with the First or Second Law.

The laws contain echoes of the Golden Rule ("Thou Shalt Not Kill"), the Judeo-Christian notion that sin results from acts committed and omitted, the physician's Hippocratic oath, and even the right to self-defense. Sounds pretty good, right? Except they never work. In "Runaround," mining engineers on the surface of Mars order a robot to retrieve an element that is poisonous to it. Instead, it gets stuck in a feedback loop between law two—obey orders—and law three—protect yourself. The robot walks in drunken circles until the engineers risk *their* lives to rescue it. And so it goes with every Asimov robot tale— unanticipated consequences result from contradictions inherent in the three laws. Only by working around the laws are disasters averted.

Asimov was generating plot lines, not trying to solve safety issues in the real world. Where you and I live his laws fall short. For starters, they're insufficiently precise. What exactly will constitute a "robot" when humans augment their bodies and brains with intelligent prosthetics and implants? For that matter, what will constitute a human? "Orders," "injure," and "existence" are similarly nebulous terms.

Tricking robots into performing criminal acts would be simple, unless the robots had perfect comprehension of all of human knowledge. "Put a little dimethylmercury in Charlie's shampoo" is a recipe for murder only if you know that dimethylmercury is a neurotoxin. Asimov eventually added a fourth law, the Zeroth Law, prohibiting robots from harming mankind as a whole, but it doesn't solve the problems.

Yet unreliable as Asimov's laws are, they're our most often cited attempt to codify our future relationship with intelligent machines. That's a frightening proposition. Are Asimov's laws all we've got?

I'm afraid it's worse than that. Semiautonomous robotic drones already kill dozens of people each year. Fifty-six countries have or are developing battlefield robots. The race is on to make them autonomous and intelligent. For the most part, discussions of ethics in AI and technological advances take place in different worlds.

As I'll argue, AI is a dual-use technology like nuclear fission. Nuclear fission can illuminate cities or incinerate them. Its terrible power was unimaginable to most people before 1945. With advanced AI, we're in the 1930s right now. We're unlikely to survive an introduction as abrupt as nuclear fission's.

Chapter Two

The Two-Minute Problem

Our approach to existential risks cannot be one of trial-and-error. There is no opportunity to learn from errors. The reactive approach—see what happens, limit damages, and learn from experience—is unworkable.

— Nick Bostrom, faculty of Philosophy, Oxford University

The AI does not hate you, nor does it love you, but you are made out of atoms which it can use for something else.

— Eliezer Yudkowsky, research fellow,
Machine Intelligence Research Institute

Artificial superintelligence does not yet exist, nor does artificial general intelligence, the kind that can learn like we do and will in many senses match and exceed most human intelligence. However, regular old artificial intelligence surrounds us, performing hundreds of tasks humans delight in having it perform. Sometimes called weak or narrow AI, it delivers remarkably useful searches (Google), suggests books you might like to read based on your prior choices (Amazon), and performs 50 to 70 percent of the buying and selling on the NYSE and the NAS-DAQ stock exchange. Because they do just one thing, albeit extremely well, heavy hitters like IBM's chess-playing Deep Blue and *Jeopardy!*-playing Watson also get squeezed into the category of narrow AI.

So far, AI has been highly rewarding. In one of my car's dozen or so computer chips, the algorithm that translates my foot pressure into an effective braking cadence (antilock braking system, or ABS) is far better at avoiding skidding than I am. Google Search has become my virtual assistant, and probably yours too. Life seems better where AI assists. And it could soon be much more. Imagine teams of a hundred Ph.D.-equivalent computers working 24/7 on important issues like cancer, pharmaceutical research and development, life extension, synthetic fuels, and climate change. Imagine the revolution in robotics, as intelligent, adaptive machines take on dangerous jobs like mining, firefighting, soldiering, and exploring sea and space. For the moment, forget the perils of self-improving superintelligence. AGI would be mankind's most important and beneficial invention.

But what exactly are we talking about when we talk about the magical quality of these inventions, their human-level *intelligence*? What does our intelligence let us humans do that other animals cannot?

Well, with your human-level smarts you can talk on the phone. You can drive a car. You can identify thousands of common objects and describe their textures and how to manipulate them. You can peruse the Internet. You may be able to count to ten in several languages, perhaps even speak fluently in more than one. You've got good commonsense knowledge—you know that handles go on doors *and* cups, and innumerable other useful facts about your environment. And you can frequently change environments, adapting to each appropriately.

You can do things in succession or in combination, or keep some in the background while focusing your attention on what's most important now. And you can effortlessly switch among the different tasks, with their different inputs, without hesitation. Perhaps most important, you can learn new skills, new facts, and plan your own self-improvement. The vast majority of living things are born with all the abilities they'll ever use. Not you.

Your remarkable gamut of high-level abilities are what we mean by human-level intelligence, the general intelligence that AGI developers seek to achieve in a machine.

Does a generally intelligent machine require a body? To meet our definition of general intelligence a computer would need ways to receive input from the environment, and provide output, but not a lot more. It needs ways to manipulate objects in the real world. But as we saw in the Busy Child scenario, a sufficiently advanced intelligence can get someone or something else to manipulate objects in the real world. Alan Turing devised a test for human-level intelligence, now called the Turing test, which we will explore later. His standard for demonstrating human-level intelligence called only for the most basic keyboard-and-monitor kind of input and output devices.

The strongest argument for why advanced AI needs a body may come from its learning and development phase—scientists may discover it's not possible to "grow" AGI without some kind of body. We'll explore the important question of "embodied" intelligence later on, but let's get back to our definition. For the time being it's enough to say that by general intelligence we mean *the ability to solve problems, learn, and take effective, human-like action, in a variety of environments.*

Robots, meanwhile, have their own row to hoe. So far, none are particularly intelligent even in a narrow sense, and few have more than a crude ability to get around and manipulate objects autonomously. Robots will only be as good as the intelligence that controls them.

Now, how long until we reach AGI? A few AI experts I've communicated with don't think 2025 is too soon to anticipate human-level artificial intelligence. But overall, recent polls show that computer scientists and professionals in AI-related fields, such as engineering, robotics, and neuroscience, are more conservative. They think there's a better than 10 percent chance AGI will be created before 2028, and a better than 50 percent chance by 2050. Before the end of this century, a 90 percent chance.

Furthermore, experts claim, the military or large businesses will achieve AGI first; academia and small organizations are less likely to. About the pros and cons, the results aren't surprising—working toward AGI will reward us with enormous benefits, and threaten us with huge disasters, including the kind from which human beings won't recover.

The greatest disasters, as we explored in chapter 1, come after the bridge from AGI—human-level intelligence—to ASI—superintelligence. And the time gap between AGI and ASI could be brief. But remarkably, while the risks involved with sharing our planet with superintelligent AI strike many in the AI community as the subject of the most important conversation anywhere, it's been all but left out of the public dialogue. Why?

There are several reasons. Most dialogues about dangerous AI aren't very broad or deep, and not many people understand them. The issues are well developed in pockets of Silicon Valley and academia, but they aren't absorbed elsewhere, most alarmingly in the field of technology journalism. When a dystopian viewpoint rears its head, many bloggers, editorialists, and technologists reflexively fend it off with some version of "Oh no, not the Terminator again! Haven't we heard enough gloom and doom from Luddites and pessimists?" This reaction is plain lazy, and it shows in flimsy critiques. The inconvenient facts of AI risk are not as sexy or accessible as techno-journalism's usual fare of dual core 3-D processors, capacitive touch screens, and the current hit app.

I also think its popularity as entertainment has inoculated AI from serious consideration in the not-so-entertaining category of catastrophic risks. For decades, getting wiped out by artificial intelligence, usually in the form of humanoid robots, or in the most artful case a glowing red lens, has been a staple of popular movies, science-fiction novels, and video games. Imagine if the Centers for Disease Control issued a serious warning about vampires (unlike their recent tongue-in-cheek alert about zombies). Because vampires have provided so much fun, it'd take time for the guffawing to stop, and the wooden stakes to come out. Maybe we're in that period right now with AI, and only an accident or a near-death experience will jar us awake.

Another reason AI and human extinction do not often receive serious consideration may be due to one of our psychological blind spots—a cognitive bias. Cognitive biases are open manholes on the avenues of our thinking. Israeli American psychologists Amos Tversky and Daniel Kahneman began developing the science of cognitive biases in 1972. Their basic idea is that we humans make decisions in irrational ways. That observation alone won't earn you a Nobel Prize

(Kahneman received one in 2002); the stunner is that we are irrational in scientifically verifiable patterns. In order to make the quick decisions useful during our evolution, we repeatedly take the same mental shortcuts, called heuristics. One is to draw broad inferences—too broad as it turns out—from our own experiences.

Say, for example, you're visiting a friend and his house catches on fire. You escape, and the next day you take part in a poll ranking causes of accidental death. Who would blame you if you ranked "fire" as the first or second most common cause? In fact, in the United States, fire ranks well down the list, after falls, traffic accidents, and poisonings. But by choosing fire, you have demonstrated what's called the "availability" bias: your recent experience impacts your decision, making it irrational. But don't feel bad—it happens to everyone, and there are a dozen more biases in addition to availability.

Perhaps it's the availability bias that keeps us from associating artificial intelligence with human annihilation. We haven't experienced well-publicized accidents at the hands of AI, while we've come close with the other usual suspects. We know about superviruses like HIV, SARS, and the 1918 Spanish Flu. We've seen the effects of nuclear weapons on cities full of humans. We've been scared by geological evidence of ancient asteroids the size of Texas. And disasters at Three Mile Island (1979), Chernobyl (1986), and Fukushima (2011) show us we must learn even the most painful lessons again and again.

Artificial intelligence is not yet on our existential threat radar. Again, an accident would change that, just as 9/11 introduced the world to the concept that airplanes could be wielded as weapons. That attack revolutionized airline security and spawned a new forty-four-billion-dollar-a-year bureaucracy, the

Department of Homeland Security. Must we have an AI disaster to learn a similarly excruciating lesson? Hopefully not, because there's one big problem with AI disasters. They're not like airplane disasters, nuclear disasters, or any other kind of technology disaster with the possible exception of nanotechnology. That's because there's a high probability we won't recover from the first one.

And there's another critical way in which runaway AI is different from other technological accidents. Nuclear plants and airplanes are one-shot affairs—when the disaster is over you clean it up. A true AI disaster involves smart software that improves itself and reproduces at high speeds. It's self-perpetuating. How can we stop a disaster if it outmatches our strongest defense—our brains? And how can we clean up a disaster that, once it starts, may never stop?

Another reason for the curious absence of AI in discussions of existential threats is that the Singularity dominates AI dialogue.

"Singularity" has become a very popular word to throw around, even though it has several definitions that are often used interchangeably. Accomplished inventor, author, and Singularity pitchman Ray Kurzweil defines the Singularity as a "singular" period in time (beginning around the year 2045) after which the pace of technological change will irreversibly transform human life. Most intelligence will be computer-based, and trillions of times more powerful than today. The Singularity will jump-start a new era in mankind's history in which most of our problems, such as hunger, disease, even mortality, will be solved.

Artificial intelligence is the star of the Singularity media spectacle, but nanotechnology plays an important supporting role. Many experts predict that artificial superintelligence will put nanotechnology on the fast track by finding solutions for seemingly intractable problems with nanotech's development. Some think it would be better if ASI came first, because nanotechnology is too volatile a tool to trust to our puny brains. In fact, a lot of the benefits that are attributed to the Singularity are due to nanotechnology, not artificial intelligence. Engineering at an atomic scale may provide, among other things: immortality, by eliminating on the cellular level the effects of aging; immersive virtual reality, because it'll come from nanobots that take over the body's sensory inputs; and neural scanning and uploading of minds to computers.

However, say skeptics, out-of-control nano robots might endlessly reproduce themselves, turning the planet into a mass of "gray goo." The "gray goo" problem is nanotechnology's most well-known Frankenstein face. But almost no one describes an analogous problem with AI, such as the "intelligence explosion" in which the development of smarter-than-human machines sets in motion the extinction of the human race. That's one of the many downsides of the Singularity spectacle, one of many we don't hear enough about. That absence may be due to what I call the two-minute problem.

I've listened to dozens of scientists, inventors, and ethicists lecture about superintelligence. Most consider it inevitable, and celebrate the bounty the ASI genie will grant us. Then, often in the last two minutes of their talks, experts note that if AI's not properly managed, it could extinguish humanity. Then their audiences nervously chuckle, eager to get back to the good news.

Authors approach the ongoing technological revolution in one of two ways. First there are books like Kurzweil's *The Singularity*

Is Near. Their goal is to lay the theoretical groundwork for a supremely positive future. If a bad thing happened there, you would never hear about it over optimism's merry din. Jeff Stibel's *Wired for Thought* represents the second tack. It looks at the technological future through the lens of business. Stibel persuasively argues that the Internet is an increasingly well-connected brain, and Web start-ups should take this into account. Books like Stibel's try to teach entrepreneurs how to dip a net between Internet trends and consumers, and seine off buckets full of cash.

Most technology theorists and authors are missing the less rosy, third perspective, and this book aims to make up for it. The argument is that the endgame for first creating smart machines, then smarter-than-human machines, is not their integration into our lives, but their conquest of us. In the quest for AGI, researchers will create a kind of intelligence that is stronger than their own and that they cannot control or adequately understand.

We've learned what happens when technologically advanced beings run into less advanced ones: Christopher Columbus versus the Tiano, Pizzaro versus the Inca, Europeans versus Native Americans.

Get ready for the next one. Artificial superintelligence versus you and me.

Perhaps technology thinkers have considered AI's downside, but believe it's too unlikely to worry about. Or they get it, but think they can't do anything to change it. Noted AI developer Ben Goertzel, whose road map to AGI we'll explore in chapter 11, told me that we won't know how to protect ourselves from advanced AI until we have had a lot more experience with it.

Kurzweil, whose theories we'll investigate in chapter 9, has long argued a similar point—our invention and integration with superintelligence will be gradual enough for us to learn as we go. Both argue that the *actual* dangers of AI cannot be seen from here. In other words, if you are living in the horse-and-buggy age, it's impossible to anticipate how to steer an automobile over icy roads. So, relax, we'll figure it out when we get there.

My problem with the gradualist view is that while superintelligent machines can certainly wipe out humankind, or make us irrelevant, I think there is also plenty to fear from the AIs we will encounter on the developmental path to superintelligence. That is, a mother grizzly may be highly disruptive to a picnic, but don't discount a juvenile bear's ability to shake things up, too. Moreover, gradualists think that from the platform of human-level intelligence, the jump to superintelligence may take years or decades longer. That would give us a grace period of coexistence with smart machines during which we could learn a lot about how to interact with them. Then their advanced descendants won't catch us unawares.

But it ain't necessarily so. The jump from human-level intelligence to superintelligence, through a positive feedback loop of self-improvement, could undergo what is called a "hard takeoff." In this scenario, an AGI improves its intelligence so rapidly that it becomes superintelligent in weeks, days, or even hours, instead of months or years. Chapter 1 outlines a hard takeoff's likely speed and impact. There may be nothing gradual about it.

It may be that Goertzel and Kurzweil are right—we'll take a closer look at the gradualist argument later. But what I want to get across right now are some important, alarming ideas derived from the Busy Child scenario.

Computer scientists, especially those who work for defense

and intelligence agencies, will feel compelled to speed up the development of AGI because to them the alternatives (such as the Chinese government developing it first) are more frightening than hastily developing their own AGI. Computer scientists may also feel compelled to speed up the development of AGI in order to better control other highly volatile technologies likely to emerge in this century, such as nanotechnology. They may not stop to consider checks to self-improvement. A self-improving artificial intelligence could jump quickly from AGI to ASI in a hard takeoff version of an "intelligence explosion."

Because we cannot know what an intelligence smarter than our own will do, we can only imagine a fraction of the abilities it may use against us, such as duplicating itself to bring more superintelligent minds to bear on problems, simultaneously working on many strategic issues related to its escape and survival, and acting outside the rules of honesty or fairness. Finally, we'd be prudent to assume that the first ASI will not be friendly or unfriendly, but ambivalent about our happiness, health, and survival.

Can we calculate the potential risk from ASI? In his book *Technological Risk*, H. W. Lewis identifies categories of risk and ranks them by how easy they are to factor. Easiest are actions of high probability and high consequence, like driving a car from one city to another. There's plenty of data to consult. Low probability, high consequence events, like earthquakes, are rarer, and therefore harder to anticipate. But their consequences are so severe that calculating their likelihood is worthwhile.

Then there are risks whose probability is low because they've never happened before, yet their consequences are, again, severe. Major climate change resulting from man-made pollution is one good example. Before the July 16, 1945, test at White

Sands, New Mexico, the detonation of an atomic bomb was another. Technically, it is in this category that superintelligence resides. Experience doesn't provide much guidance. You cannot calculate its probability using traditional statistical methods.

I believe, however, that given the current pace of AI development the invention of superintelligence belongs in the first category—a high probability and high-risk event. Furthermore, even if it were a low probability event, its risk factor should promote it to the front tier of our attention.

Put another way, I believe the Busy Child will come very soon.

The fear of being outsmarted by greater-than-human intelligence is an old one, but early in this century a sophisticated experiment about it came out of Silicon Valley, and instantly became the stuff of Internet legend.

The rumor went like this: a lone genius had engaged in a series of high-stakes bets in a scenario he called the AI-Box Experiment. In the experiment, the genius role-played the part of the AI. An assortment of dot-com millionaires each took a turn as the Gatekeeper—an AI maker confronted with the dilemma of guarding and containing smarter-than-human AI. The AI and Gatekeeper would communicate through an online chat room. Using only a keyboard, it was said, the man posing as the ASI escaped every time, and won each bet. More important, he proved his point. If he, a mere human, could talk his way out of the box, an ASI hundreds or thousands of times smarter could do it too, and do it much faster. This would lead to mankind's likely annihilation.

The rumor said the genius had gone underground. He'd garnered so much notoriety for the AI-Box Experiment, and for authoring papers and essays on AI, that he had developed a fan base. Spending time with fans was less rewarding than the reason he'd started the AI-Box Experiment to begin with—to save mankind.

Therefore, he had made himself hard to find. But of course I wanted to talk to him.

Chapter Three

Looking into the Future

AGI is intrinsically very, very dangerous. And this problem is not terribly difficult to understand. You don't need to be super smart or super well informed, or even super intellectually honest to understand this problem.

—Michael Vassar, president,
Machine Intelligence Research Institute

"I definitely think that people should try to develop Artificial General Intelligence with all due care. In this case, all due care means much more scrupulous caution than would be necessary for dealing with Ebola or plutonium."

Michael Vassar is a trim, compact man of about thirty. He holds degrees in biochemistry and business, and is fluent in assessments of human annihilation, so words like "Ebola" and "plutonium" come out of his mouth without hesitation or irony. One wall of his high-rise condo is a floor-to-ceiling window, and it frames a red suspension bridge that links San Francisco to Oakland, California. This isn't the elegant Golden Gate—that's across town. This one has been called its ugly stepsister. Vassar told me people bent on committing suicide have been known to drive *over* this bridge to get to the nice one.

Vassar has devoted his life to thwarting suicide on a larger scale. He's the president of the Machine Intelligence Research Institute, a San Francisco–based think tank established to fight the extinction of the human race at the hands, or bytes, of artificial intelligence. On its Web site, MIRI posts thoughtful papers on dangerous aspects of AI, and once a year it organizes the influential Singularity Summit. At the two-day conference, programmers, neuroscientists, academics, entrepreneurs, ethicists, and inventors hash out advances and setbacks in the ongoing AI revolution. MIRI invites talks from believers and nonbelievers alike, people who don't think the Singularity will ever happen, and people who think MIRI is an apocalyptic techno cult.

Vassar smiled at the cult idea. "People who come to work for MIRI are the opposite of joiners. Usually they realize AI's dangers before they even know MIRI exists."

I didn't know MIRI existed until after I'd heard about the AI-Box Experiment. A friend had told me about it, but in the telling he got a lot wrong about the lone genius and his millionaire opponents. I tracked the story to a MIRI Web site, and discovered that the experiment's creator, Eliezer Yudkowsky, had cofounded MIRI (then called the Singularity Institute for Artificial Intelligence) with entrepreneurs Brian and Sabine Atkins. Despite his reputed reticence, Yudkowsky and I exchanged e-mails and he gave me the straight dope about the experiment.

The bets placed between the AI played by Yudkowsky and the Gatekeeper assigned to rein him in were at most thousands of dollars, not millions. The game had been held just five times, and the AI in the box won three of these times. Meaning, the AI usually got out of the box, but it wasn't a blowout.

Some parts of the AI-Box rumor had been true—Yudkowsky *was* reclusive, stingy with his time, and secretive about where he lived. I had invited myself to Michael Vassar's home because I was pleased and amazed that a nonprofit had been founded to combat the dangers of AI, and young, intelligent people were devoting their lives to the problem. And I hoped my conversation with Vassar would smooth my final steps to Yudkowsky's front door.

Before jumping feet first into AI danger advocacy, Vassar had earned an MBA and made money cofounding Sir Groovy, an online music-licensing firm. Sir Groovy pairs independent music labels with TV and film producers to provide fresh soundtracks from lesser known and hence cheaper artists. Vassar had been toying with the idea of applying himself to the dangers of nanotechnology until 2003. That year he met Eliezer Yudkowsky, after having read his work online for years. He learned about MIRI, and a threat more imminent and dangerous than nanotechnology: artificial intelligence.

"I became extremely concerned about global catastrophic risk from AGI after Eliezer convinced me that it was plausible that AGI could be developed in a short time frame and on a relatively small budget. I didn't have any convincing reason to think that AGI could *not* happen say in the next twenty years." That was sooner than predictions for nanotech. And AGI's development would take a lot less overhead. So Vassar changed course.

When we met, I confessed I hadn't thought much about the idea that small groups with small budgets could come up with

AGI. From the polls I'd seen, only a minority of experts predicted such a team would be the likely parents.

So, could Al Qaeda create AGI? Could FARC? Or Aum Shinrikyo?

Vassar doesn't think a terrorist cell will come up with AGI. There's an IQ gap.

"The bad guys who actually want to destroy the world are reliably not very capable. You know the sorts of people who do want to destroy the world lack the long-term planning abilities to execute anything."

But what about Al Qaeda? Didn't all the attacks up to and including 9/11 require high levels of imagination and planning?

"They do not compare to creating AGI. Writing code for an application that does any one thing better than a human, never mind the panoply of capabilities of AGI, would require orders of magnitude more talent and organization than demonstrated by Al Qaeda's entire catalogue of violence. If AGI were that easy, someone smarter than Al Qaeda would have already done it."

But what about governments like those of North Korea and Iran?

"As a practical matter the quality of science that bad regimes produce is shit. The Nazis are the only exception and, well, if the Nazis happen again we have very big problems with or without AI."

I disagreed, though not about the Nazis. Iran and North Korea have found high-tech ways to blackmail the rest of the world with the development of nuclear weapons and intercontinental missiles. So I wouldn't cross them off the short list of potential AGI makers with a track record of blowing raspberries in the face of international censure. Plus, if AGI can be created by small groups, any rogue state could sponsor one.

When Vassar talked about small groups, he included companies working under the radar. I'd heard about so-called stealth companies that are privately held, hire secretly, never issue press releases or otherwise reveal what they're up to. In AI, the only reason for a company to be stealthy is if they've had some powerful insight, and they don't want to reward competitors with information about what that their breakthrough is. By definition, stealth companies are hard to discover, though rumors abound. PayPal founder Peter Thiel funds three stealth companies devoted to AI.

Companies in "stealth mode" however, are different and more common. These companies seek funding and even publicity, but don't reveal their plans. Peter Voss, an AI innovator known for developing voice-recognition technology, pursues AGI with his company, Adaptive AI, Inc. He has gone on record saying AGI can be achieved within ten years. But he won't say how.

Stealth companies come with another complication. A small, motivated company could exist within a larger company with a big public presence. What about Google? Why wouldn't the cash-rich megacorp take on AI's Holy Grail?

When I questioned him at an AGI conference, Google's Director of Research Peter Norvig, coauthor of the classic AI textbook, *Artificial Intelligence: A Modern Approach*, said Google wasn't looking into AGI. He compared the quest to NASA's plan for manned interplanetary travel. It doesn't have one. But it will continue to develop the component sciences of traveling in space—rocketry, robotics, astronomy, et cetera—and one day all the pieces will come together, and a shot at Mars will look feasible.

Likewise, narrow AI projects do lots of intelligent jobs like search, voice recognition, natural language processing, visual perception, data mining, and much more. Separately they are well-funded, powerful tools, dramatically improving each year. Together they advance the computer sciences that will benefit AGI systems.

However, Norvig told me, no AGI program for Google exists. But compare that statement to what his boss, Google cofounder Larry Page said at a London conference called Zeitgeist '06:

People always make the assumption that we're done with search. That's very far from the case. We're probably only 5 percent of the way there. We want to create the ultimate search engine that can understand anything . . . some people could call that artificial intelligence. . . . The ultimate search engine would understand everything in the world. It would understand everything that you asked it and give you back the exact right thing instantly. . . . You could ask "what should I ask Larry?" and it would tell you.

That sounds like AGI to me.

Through several well-funded projects, IBM pursues AGI, and DARPA seems to be backing every AGI project I look into. So, again, why not Google? When I asked Jason Freidenfelds, from Google PR, he wrote:

. . . it's much too early for us to speculate about topics this far down the road. We're generally more focused on practical machine learning technologies like machine vision, speech recognition, and machine translation, which essentially is about building statistical models to match patterns—nothing close to the "thinking machine" vision of AGI.

But I think Page's quotation sheds more light on Google's attitudes than Freidenfelds's. And it helps explain Google's evolution from the visionary, insurrectionist company of the 1990s, with the much touted slogan DON'T BE EVIL, to today's opaque, Orwellian, personal-data-aggregating behemoth.

The company's privacy policy shares your personal information among Google services, including Gmail, Google+, YouTube, and others. Who you know, where you go, what you buy, who you meet, how you browse—Google collates it all. Its purported goal: to improve your user experience by making search virtually omniscient about the subject of *you*. It's parallel goal—to shape what ads you see, and even your news, videos, and music consumption, and automatically target you with marketing campaigns. Even the Google camera cars that take "Street View" photographs for Google Maps are part of the plan—for three years, Google used its photo-taking fleet to grab data from private Wi-Fi networks in the United States and elsewhere. Passwords, Internet usage history, personal e-mails—nothing was off limits.

It's clear they've put once loyal customers in our place, and it's not first place. So it seemed inconceivable that Google did not have AGI in mind.

Then, about a month after my last correspondence with Freidenfelds, *The New York Times* broke a story about Google X.

Google X was a stealth company. The secret Silicon Valley laboratory was initially headed by AI expert and developer of

Google's self-driving car, Sebastian Thrun. It is focused on one hundred "moon-shot" projects such as the Space Elevator, which is essentially a scaffolding that would reach into space and facilitate the exploration of our solar system. Also onboard at the stealth facility is Andrew Ng, former director of Stanford University's Artificial Intelligence Lab, and a world-class roboticist.

Finally, late in 2012, Google hired esteemed inventor and author Ray Kurzweil to be its director of engineering. As we'll discuss in chapter 9, Kurzweil has a long track record of achievements in AI, and has promoted brain research as the most direct route to achieving AGI.

It doesn't take Google glasses to see that if Google employs at least two of the world's preeminent AI scientists, and Ray Kurzweil, AGI likely ranks high among its moon-shot pursuits.

Seeking a competitive advantage in the marketplace, Google X and other stealth companies may come up with AGI away from public view.

Stealth companies may represent a surprise track to AGI. But according to Vassar the quickest path to AGI will be very public, and cost serious money. That route calls for reverse engineering the human brain, using a combination of programming skill and brute force technology. "Brute force" is the term for overpowering a problem with sheer hardware muscle—racks of fast processors, petabytes of memory—along with clever programming.

"The extreme version of brute force is coming out of biology," Vassar told me. "If people continue to use machines to analyze biological systems, work out metabolisms, work out these complex relationships inside biology, eventually they'll accumulate a lot of information on how neurons process information. And once they have enough information about how neurons process information, that information can be analyzed for AGI purposes."

It works like this: *thinking* runs on biochemical processes performed by parts of the brain called neurons, synapses, and dendrites. With a variety of techniques, including PET and fMRI brain scanning, and applying neural probes inside and outside the skull, researchers determine what individual neurons and clusters of neurons are doing in a computational sense. Then they express each of these processes with a computer program or algorithm.

That's the thrust of the new field of computational neuroscience. One of the field's leaders, Dr. Richard Granger, director of the Dartmouth University Brain Engineering Laboratory, has created algorithms that mimic circuits in the human brain. He's even patented a hugely powerful computer processor based on how these brain circuits work. When it gets to market we'll see a giant leap forward in how computer systems visually identify objects because they'll do it the way our brain does.

There are still many brain circuits remaining to probe and map. But once you've created algorithms for all the brain's processes, congratulations, you have a brain. Or do you? Maybe not. Maybe what you have is a machine that emulates a brain. This is a big question in AI. For instance, does a chess-playing program think?

When IBM set out to create Deep Blue, and defeat the world's best chess players, they didn't program it to play chess like World Champion Gary Kasparov only better. They didn't know how. Kasparov developed his virtuosity by playing a vast number of games, and studying more games. He developed a huge repository of opens, attacks, feints, blockades, decoys, gambits, endgames—strategies and tactics. He recognizes board patterns, remembers, and *thinks*. Kasparov normally thinks three to five moves ahead, but can go as far as fourteen. No current computer can do all that.

So instead, IBM programmed a computer to evaluate 200 million positions per second.

First Deep Blue would make a hypothetical move, and evaluate all of Kasparov's possible responses. It would make *its* hypothetical response to each of those responses, and again evaluate all of Kasparov's responses. This two-levels-deep modeling is called a two-ply search—Deep Blue would sometimes search up to six plies deep. That's each side "moving" six times for every hypothetical move.

Then Deep Blue would go back to the still untouched board, and begin evaluating another move. It would repeat this process for many possible moves all while scoring each move according to whether it captured a piece, the value of the piece, whether it improved its overall board position, and by how much. Finally, it would play the highest scoring move.

Was Deep Blue thinking?

Maybe. But few would argue it was thinking the way a human thinks. And few experts doubt that it'll be the same way with AGI. Each researcher trying to achieve AGI has their own approach. Some are purely biological, working to closely mimic the brain. Others are biologically inspired, taking the brain as a cue, but relying more on AI's hardworking tool kit: theorem provers, search algorithms, learning algorithms, automated reasoning, and more.

We'll get into some of these, and explore how the human brain actually uses many of the same computation techniques as computers. But the point is, it's not clear if computers will think as we define it, or if they'll ever possess anything like intention or consciousness. Therefore, some scholars say, artificial intelligence equivalent to human intelligence is impossible.

Philosopher John Searle created a thought experiment called the Chinese Room Argument that aims to prove this point:

> Imagine a native English speaker who knows no Chinese locked in a room full of boxes of Chinese symbols (a data base) together with a book of instructions for manipulating the symbols (the program). Imagine that people outside the room send in other Chinese symbols, which, unknown to the person in the room, are questions in Chinese (the input). And imagine that by following the instructions in the program the man in the room is able to pass out Chinese symbols, which are correct answers to the questions (the output).

The man inside the room answers correctly, so the people outside think he can communicate in Chinese. Yet the man doesn't understand a word of Chinese. Like the man, Searle concludes, a computer will never really think or understand. At best what researchers will get from efforts to reverse engineer the brain will be a refined mimic. And AGI systems will achieve similarly mechanical results.

Searle's not alone in believing computers will never think or attain consciousness. But he has many critics, with many different complaints. Some detractors claim he is computerphobic. Taken as a whole, everything in the Chinese room, including the man, come together to create a system that persuasively "understands" Chinese. Seen this way Searle's argument is circular: no part of the room (computer) understands Chinese, ergo the computer cannot understand Chinese.

And you can just as easily apply Searle's objection to humans: we don't have a formal description of what understanding language really is, so how can we claim humans "understand" language? We have only observation to confirm that language is understood. Just like the people outside Searle's room.

What's so remarkable about the brain's processes, even consciousness, anyway? Just because we don't understand consciousness now doesn't mean we never will. It's not magic.

Still, I agree with Searle *and* his critics. Searle is correct in thinking AGI won't be like us. It will be full of computational techniques whose operation no one fully understands. And computer systems designed to create AGI, called "cognitive architectures," may be too complex for any one person to grasp anyway. But Searle's critics are correct in thinking that someday an AGI or ASI *could* think like us, if we make it that far.

I don't believe we will. I think our Waterloo lies in the foreseeable future, in the AI of tomorrow and the nascent AGI due out in the next decade or two. Our survival, if it is possible, may depend on, among other things, developing AGI with something akin to consciousness and human understanding, even friendliness, built in. That would require, at minimum, understanding intelligent machines in a fine-grained way, so there'd be no surprises.

Let's go back to one common definition of the Singularity for a moment, what's called the "technological Singularity." It refers to the time in history when we humans share the planet with smarter-than-human intelligence. Ray Kurzweil proposes that we'll merge with the machines, ensuring our survival. Others propose the machines will enhance our lives, but we'll continue to live as regular old humans, not human-machine cyborgs. Still others, like me, think the future belongs to machines.

The Machine Intelligence Research Institute was formed to ensure that whatever form our heirs take, our values will be preserved.

In his San Francisco high-rise apartment, Vassar told me, "The stakes are the delivery of human value to humanity's successors. And through them to the universe."

To MIRI, the first AGI out of the box must be safe, and therefore carry human value to humanity's successors in whatever form they appear. If the AGI is not safe, neither humans nor what we value will survive. And it's not just the future of the earth that's on the block. As Vassar told me, "MIRI's mission is to cause the technological singularity to happen in the best possible way, to bring about the best possible future for the universe."

What would a good outcome for the universe look like?

Vassar gazed out the window at the rush-hour traffic that was just starting to stack up on the iron bridge to Oakland. Somewhere beyond the water lay the future. In his mind, superintelligence has already escaped us. It has colonized our solar system, then our galaxy. Now it was reformatting the universe with megascale building projects, and growing into something so unusual it's hard for us to grasp.

In that future, he told me, the entire universe becomes a computer or mind, as far beyond our ken as spaceships are to flatworms. Kurzweil writes that this is the universe's destiny. Others agree, but believe that with the reckless development of advanced AI we'll assure our elimination as well as that of other beings that might be out there. Just as ASI may not hate us or love us, neither will it hate or love other creatures in the universe. Is our quest for AGI the start of a galaxy-wide plague?

As I left Vassar's apartment I wondered what could prevent this dystopian vision from coming true. What could stop the annihilating kind of AGI? Furthermore, were there holes in the dystopian hypothesis?

Well, builders of AI and AGI could make it "friendly," so that whatever evolves from the first AGI won't destroy us and other creatures in the universe. Or, we might be wrong about AGI's abilities and "drives," and fearing its conquest of the universe could be a false dilemma.

Maybe AI can never advance to AGI and beyond, or maybe there are good reasons to think it will happen in a different and more manageable way than we currently think possible. In short, I wanted to know what could put us on a safer course to the future.

I intended to ask the AI Box Experiment creator, Eliezer Yudkowsky. Besides originating that thought experiment, I'd been told that he knew more about Friendly AI than anyone else in the world.

Chapter Four

The Hard Way

With the possible exception of nanotechnology being released upon the world there is nothing in the whole catalogue of disasters that is comparable to AGI.

—Eliezer Yudkowsky, Research Fellow,
Machine Intelligence Research Institute

Fourteen "official" cities comprise Silicon Valley, and twenty-five math-and-engineering-focused universities and extension campuses inhabit them. They feed the software, semiconductor, and Internet firms that are the latest phase of a technology juggernaut that began here with radio in the first part of the twentieth century. Silicon Valley attracts a third of all the venture capital in the United States. It has the highest number of technology workers per capita of any U.S. metropolitan area, and they're the best paid, too. The country's greatest concentration of billionaires and millionaires call Silicon Valley home.

Here, at the epicenter of global technology, with a GPS in

my rental car and another in my iPhone, I drove to Eliezer Yudkowsky's home the old-fashioned way, with written directions. To protect his privacy, Yudkowsky had e-mailed them to me and asked me not to share them or his e-mail address. He didn't offer his phone number.

At thirty-three, Yudkowsky, cofounder and research fellow at MIRI, has written more about the dangers of AI than anyone else. When he set out on this career more than a decade ago, he was one of very few people who had made considering AI's dangers his life's work. And while he hasn't taken actual vows, he forgoes activities that might take his eye off the ball. He doesn't drink, smoke, or do drugs. He rarely socializes. He gave up reading for fun several years ago. He doesn't like interviews, and prefers to do them on Skype with a thirty-minute time limit. He's an atheist (the rule not the exception among AI experts) so he doesn't squander hours at a temple or a church. He doesn't have children, though he's fond of them, and thinks people who haven't signed their children up for cryonics are lousy parents.

But here's the paradox. For someone who supposedly treasures his privacy, Yudkowsky has laid bare his personal life on the Internet. I found, after my first attempts to track him down, that in the corner of the Web where discussions of rationality theory and catastrophe live, he and his innermost musings are unavoidable.

His ubiquity is how I came to know that at age nineteen, in their hometown of Chicago, his younger brother, Yehuda, killed himself. Yudkowsky's grief came out in an online rant that still seems raw almost a decade later. And I learned that since dropping out of school in eighth grade, he has taught himself mathematics, logic, science history, and whatever else he felt compelled to know on an "as needed" basis. The other skills he's acquired include delivering compelling talks and writing dense, often funny prose:

> I'm a great fan of Bach's music, and believe that it's best rendered as techno electronica with heavy thumping beats, the way Bach intended.

Yudkowsky is a man in a hurry, because his job comes with an expiration date: when someone creates AGI. If researchers build it with the proper Yudkowsky-inspired safeguards, he may have saved mankind and maybe more. But if an intelligence explosion kicks in and Yudkowsky has been unsuccessful in implementing safeguards, there's a good chance we'll all be goo, and so will the universe. That puts Yudkowsky at the dead center of his own cosmology.

I had come here to learn more about Friendly AI, a term he coined. According to Yudkowsky, Friendly AI is the kind that will preserve humanity and our values forever. It doesn't annihilate our species or spread into the universe like a planet-eating space plague.

But what is Friendly AI? How do you create it?

I also wanted to hear about the AI Box Experiment. I especially wanted to know, as he role-played the part of the AGI, how he talked the Gatekeeper into setting him free. Someday I expect that you, someone you know, or someone a couple of people removed from you, will be in the Gatekeeper's seat. He or she needs to know what to anticipate, and how to resist. Yudkowsky might know.

Yudkowsky's condo is an end unit in a horseshoe of two-story garden apartments with a pond and electric waterfall in the central courtyard. Inside, his apartment is spotless and airy. A PC and monitor dominate the breakfast island, where he'd planted a sole padded barstool from which he could look out onto the courtyard. From here he does his writing.

Yudkowsky is tall, nearly six feet and leaning toward endomorphism—that is, he's round, but not fat. His gentle, disarming manners were a welcome change from the curt, one-line e-mails that had been our relationship's thin thread.

We sat on facing couches. I told Yudkowsky my central fear about AGI is that there's no programming technique for something as nebulous and complex as morality, or friendliness. So we'll get a machine that'll excel in problem solving, learning, adaptive behavior, and commonsense knowledge. We'll think it's humanlike. But that will be a tragic mistake.

Yudkowsky agreed. "If the programmers are less than overwhelmingly competent and careful about how they construct the AI then I would fully expect you to get something very alien. And here's the scary part. Just like dialing nine-tenths of my phone number correctly does not connect you to someone who is 90 percent similar to me. If you are trying to construct the AI's whole system and you get it 90 percent right, the result is not 90 percent good."

In fact, it's 100 percent bad. Cars aren't out to kill you, Yudkowsky analogized, but their potential deadliness is a side effect of building cars. It would be the same with AI. It wouldn't hate you, but you are made of atoms it may have other uses for, and it would, Yudkowsky said, ". . . tend to resist anything you did to try and keep those atoms to yourself." So, a side effect of thoughtless programming is that the resulting AI will have a galling lack of propriety about your atoms.

And neither the public nor the AI's developers will see the danger coming until it's too late.

"Here is this tendency to think that well-intentioned people create nice AIs, and badly intentioned people create evil AIs. This is not the source of the problem. The source of the problem is that even when well-intentioned people set out to create AIs they are not very concerned with Friendly AI issues. They themselves assume that if they are good-intentioned people the AIs they make are automatically good intentioned, and this is not true. It's actually a very difficult mathematical and engineering problem. I think most of them are just insufficiently good at thinking of uncomfortable thoughts. They started out *not* thinking, 'Friendly AI is a problem that will kill you.' "

Yudkowsky said that AI makers are infected by the idea of a blissful AI-enhanced future that lives in their imaginations. They have been thinking about it since the AI bug first bit them.

"They do not want to hear anything that contradicts that. So if you present unfriendly AI to them it bounces off. As the old proverb goes, most of the damage is done by people who wish to feel themselves important. Many ambitious people find it far less scary to think about destroying the world than to think about never amounting to much of anything at all. *All* the people I have met who think they are going to win eternal fame through their AI projects have been like this."

These AI makers aren't mad scientists or people any different from you and me—you'll meet several in this book. But recall the availability bias from chapter 2. When faced with a decision, humans will choose the option that's recent, dramatic, or otherwise front and center. Annihilation by AI isn't generally available to AI makers. Not as available as making advances in their field, getting tenure, publishing, getting rich, and so on.

In fact, not many AI makers, in contrast to AI *theorists*, are concerned with building Friendly AI. With one exception, none of the dozen or so AI makers I've spoken with are worried enough to work on Friendly AI or any other defensive measure. Maybe the thinkers overestimate the problem, or maybe the *makers'* problem is not knowing what they don't know. In a much-read online paper, Yudkowsky put it like this:

> The human species came into existence through natural selection, which operates through the nonchance retention of chance mutations. One path leading to global catastrophe—to someone pressing the button with a mistaken idea of what the button does—is that Artificial Intelligence comes about through a similar accretion of working algorithms, with the *researchers having no deep understanding of how the combined system works.* [italics mine]
>
> Not knowing how to build a Friendly AI is not deadly, of itself. . . . It's the mistaken belief that an AI will be friendly which implies an obvious path to global catastrophe.

Assuming that human-level AIs (AGIs) will be friendly is wrong for a lot of reasons. The assumption becomes even more dangerous after the AGI's intelligence rockets past ours, and it becomes ASI—artificial superintelligence. So how do you create friendly AI? Or could you impose friendliness on advanced AIs after they're already built? Yudkowsky has written a book-length online treatise about these questions entitled *Creating Friendly AI: The Analysis and Design of Benevolent Goal Architectures.* Friendly AI is a subject so dense yet important it exasperates its chief proponent himself, who says about it, "it only takes one error for a chain of reasoning to end up in Outer Mongolia."

Let's start with a simple definition. Friendly AI is *AI that has a positive rather than a negative impact on mankind.* Friendly AI pursues goals, and it takes action to fulfill those goals. To describe an AI's success at achieving its goals, theorists use a term from economics: utility. As you might recall from Econ 101, consumers behaving rationally seek to maximize utility by spending their resources in the way that gives them the most satisfaction. Generally speaking, for an AI, satisfaction is gained by achieving goals, and an act that moves it toward achieving its goals has high "utility."

Values and preferences in addition to goal satisfaction can be packed into an AI's definition of utility, called its "utility function." Being friendly to humans is one such value we'd like AIs to have. So that no matter what an AI's goals—from playing chess to driving cars—preserving human values (and humans themselves) must be an essential part of its code.

Now, *friendly* here doesn't mean Mister Rogers–friendly, though that wouldn't hurt. It means that AI should be neither hostile nor ambivalent toward humans, *forever,* no matter what its goals are or how many self-improving iterations it goes through. The AI must have an understanding of our nature so deep that it doesn't harm us through unintended consequences, like those caused by Asimov's Three Laws of Robotics. That is, we don't want an AI that meets our short-term goals—please save us from hunger—with solutions detrimental in the long term—by roasting every chicken on earth—or with solutions to which we'd object—by killing us after our next meal.

As an example of unintended consequences, Oxford University ethicist Nick Bostrom suggests the hypothetical "paper clip maximizer." In Bostrom's scenario, a thoughtlessly programmed superintelligence whose programmed goal is to manufacture paper clips does exactly as it is told without regard to human values. It all goes wrong because it sets about "transforming first all of earth and then increasing portions of space into paper clip manufacturing facilities." Friendly AI would make only as many paper clips as was compatible with human values.

Another tenet of Friendly AI is to avoid dogmatic values. What we consider to be good changes with time, and any AI involved with human well-being will need to stay up to speed. If in its utility function an AI sought to preserve the preferences of most Europeans in 1700 and never upgraded them, in the twenty-first century it might link our happiness and welfare to archaic values like racial inequality and slaveholding, gender inequality, shoes with buckles, and worse. We don't want to lock specific values into Friendly AI. We want a moving scale that evolves with us.

Yudkowsky has devised a name for the ability to "evolve" norms—Coherent Extrapolated Volition. An AI with CEV could anticipate what we would want. And not only what we would want, but what we would want if we "knew more, thought faster, and were more the people we thought we were."

CEV would be an oracular feature of friendly AI. It would have to derive from us our values *as if* we were better versions of ourselves, and be democratic about it so that humankind is not tyrannized by the norms of a few.

Does this sound a little starry-eyed? Well, there are good reasons for that. First, I'm giving you a highly summarized account of Friendly AI and CEV, concepts you can read volumes about online. And second, the whole topic of Friendly AI is incomplete and optimistic. It's unclear whether or not Friendly AI can be expressed in a formal, mathematical sense, and so there may be no way to build it or to integrate it into promising AI architectures. But if we could, what would the future look like?

Let's say that sometime, ten to forty years from now, IBM's SyNAPSE project to reverse engineer the brain has borne fruit. Jump-started in 2008 with a nearly $30 million grant from DARPA, IBM's system copies the mammalian brain's basic technique: simultaneously taking in thousands of sources of input, evolving its core processing algorithms, and outputting perception, thought, and action. It started out as a cat-sized brain, but it scaled to human-sized, and then, beyond.

To build it, the researchers of SyNAPSE (Systems of Neuromorphic Adaptive Plastic Scalable Electronics) created a "cognitive computer" made up of thousands of parallel processing computer chips. Taking advantage of developments in nanotechnology, they built chips one square micron in size. Then they arrayed the chips in a carbon sphere the size of a basketball, and suspended it in gallium aluminum alloy, a liquid metal, for maximum conductivity.

The tank holding it, meanwhile, is a powerful wireless router connected to millions of sensors distributed around the planet, and linked to the Internet. These sensors gather input from cameras, microphones, pressure and temperature gauges, robots, and natural systems—deserts, glaciers, lakes, rivers, oceans, and rain forests. SyNAPSE processes the information by automatically learning the features and relationships revealed in the massive amounts of data. Function follows form as neuromorphic, brain-imitating hardware autonomously gives rise to intelligence.

Now SyNAPSE mirrors the human brain's thirty billion neurons and hundred trillion connecting points, or synapses. And it's surpassed the brain's approximately thousand trillion operations per second.

For the first time, the human brain is the *second*-most-complex object in the known universe.

And friendliness? Knowing that "friendliness" had to be a core part of any intelligent system, its makers encoded values and safety into each of SyNAPSE's millions of chips. It is friendly down to its DNA. Now as the cognitive computer grows more powerful it makes decisions that impact the world—how to handle the AIs of terrorist states, for example, how to divert an approaching asteroid, how to stop the sea level's rapid rise, how to speed the development of nano-medicines that will cure most diseases.

With its deep understanding of humans SyNAPSE extrapolates with ease what we would choose *if* we were powerful and intelligent enough to take part in these high-level judgments. In the future, we survive the intelligence explosion! In fact, we thrive.

God bless you, Friendly AI!

Now that most (but not all) AI makers and theorists have recognized Asimov's Three Laws of Robotics for what they were meant to be—tools for drama, not survival—Friendly AI may be the best concept humans have come up with for planning their survival. But besides not being ready yet, it's got other big problems.

First, there are too many players in the AGI sweepstakes. Too many organizations in too many countries are working on AGI and AGI-related technologies for them all to agree to mothball their projects until Friendly AI is created, or to include in their code a formal friendliness module, if one could be made. And few are even taking part in the public dialogue about the necessity for Friendly AI.

Some of the AGI contestants include: IBM (with several AGI-related projects), Numenta, AGIRI, Vicarious, Carnegie Mellon's NELL and ACT-R, SNERG, LIDA, CYC, and Google. At least a dozen more, such as SOAR, Novamente, NARS, AIXItl, and Sentience, are being developed with less certain sources of funding. Hundreds more projects wholly or partially devoted to AGI exist at home and abroad, some cloaked in stealth, and some hidden behind modern-day "iron curtains" of national security in countries such as China and Israel. DARPA publicly funds many AI-related projects, but of course it funds others covertly, too.

My point is that it's unlikely MIRI will create the first AGI out of the box with friendliness built in. And it's unlikely that the first AGI's creators will think hard about issues like friendliness. Still, there is more than one way to block *unfriendly* AGI. MIRI President Michael Vassar told me about the organization's outreach program aimed at elite universities and mathematics competitions. With a series of "rationality boot camps" MIRI and its sister organization, the Center for Applied Rationality (CFAR), hope to train tomorrow's potential AI builders and technology policy makers in the discipline of rational thinking.

When these elites grow up, they'll use that education in their work to avoid AI's most harrowing pitfalls.

Quixotic as this scheme may sound, MIRI and CFAR have their fingers on an important factor in AI risk. The Singularity is trending high, and Singularity issues will come to the attention of more and smarter people. A window for education about AI risk is starting to open. But any plan to create an advisory board or governing body over AI is already too late to avoid some kinds of disasters. As I mentioned in chapter 1, at least fifty-six countries are developing robots for the battlefield. At the height of the U.S. occupation of Iraq, three Foster-Miller SWORDS—machine gun-wielding robot "drones"—were removed from combat after they allegedly pointed their guns at "friendlies." In 2007 in South Africa, a robotic antiaircraft gun killed nine soldiers and wounded fifteen in an incident lasting *an eighth of a second*.

These aren't full-blown *Terminator* incidents, but look for more of them ahead. As advanced AI becomes available, particularly if it's paid for by DARPA and like agencies in other countries, nothing will stop it from being installed in battlefield robots. In fact, robots may be the platforms for embodied machine learning that will help create advanced AI to begin with. When Friendly AI is available, if ever, why would privately run robot-making companies install it in machines designed to kill humans? Shareholders wouldn't like that one bit.

Another problem with Friendly AI is this—how will friendliness survive an intelligence explosion? That is, how will Friendly AI stay friendly even after its IQ has grown by a thousand times? In his writing and lectures, Yudkowsky employs a pithy shorthand for describing how this could happen:

Gandhi doesn't want to kill people. If you offered Gandhi a pill that made him want to kill people, he would refuse to take it, because he knows that then he would kill people, and the current Gandhi doesn't want to kill people. This, roughly speaking, is an argument that minds sufficiently advanced to precisely modify and improve themselves, will tend to preserve the motivational framework they started in.

This didn't make sense to me. If we cannot know what a smarter-than-human intelligence will do, how can we know if it will retain its utility function, or core set of beliefs? Might not it consider and reject its programmed friendliness once it's a thousand times smarter?

"Nope," Yudkowsky replied when I asked. "It becomes a thousand times more effective in *preserving* its utility function."

But what if there is some kind of category shift once something becomes a thousand times smarter than we are, and we just can't see it from here? For example, we share a lot of DNA with flatworms. But would we be invested in their goals and morals even if we discovered that many millions of years ago flatworms had created us, and given us their values? After we got over the initial surprise, wouldn't we just do whatever we wanted?

"It's very clear why one would be suspicious of that," Yudkowsky said. "But creating Friendly AI is not like giving instructions to a human. Humans have their own goals already, they have their own emotions, they have their own enforcers. They have their own structure for reasoning about moral beliefs. There is something inside that looks over any instruction you give them and decides whether to accept it or reject it. With the AI you are shaping the entire mind from scratch. If you subtract the AI's code what you are left with is a computer that is not doing anything because it has no code to run."

Still, I said, "If tomorrow I were a thousand times smarter than today, I think I'd look back at what I was worried about today and be 'so over that.' I can't believe that much of what I valued yesterday would matter to my new thousand-power mind."

"You have got a specific 'so over that' emotion and you're assuming that super intelligence would have it too," Yudkowsky said. "That is *anthropomorphism*. AI does not work like you do. It does not have a 'so over that' emotion."

But, he said, there is one exception. Human minds uploaded into computers. That's another route to AGI and beyond, sometimes confused with reverse engineering the brain. Reverse engineering seeks to first complete fine-grained learning about the human brain, then represent what the brain does in hardware and software. At the end of the process you have a computer with human-level intelligence. IBM's Blue Brain project intends to accomplish this by the early 2020s.

On the other hand, mind-uploading, also called whole brain emulation, is the theory of modeling a human mind, like yours, in a computer. At the end of the process you still have your brain (unless, as experts warn, the scanning and transfer process destroys it) but another thinking, feeling "you" exists in the machine.

"If you had a superintelligence that started out as a human upload and began improving itself and became more and more alien over time, that might turn against humanity for reasons roughly analogous to the ones that you are thinking of," Yudkowsky said. "But for a nonhuman-derived synthesized AI to turn on you, that can never happen because it is more alien than that. The vast majority of them would still kill you but not for that. Your whole visualization would apply only to a super-intelligence that came from human stock."

I'd find in my ongoing inquiry that lots of experts took issue with Friendly AI, for reasons different from mine. The day after meeting Yudkowsky I got on the phone with Dr. James Hughes, chairman of the Department of Philosophy at Trinity College, and the executive director of the Institute for Ethics and Emerging Technologies (IEET). Hughes probed a weakness in the idea that an AI's utility function couldn't change.

"One of the dogmas of the Friendly AI people is that if you are careful you can design a superintelligent being with a goal set that will become unchanging. And they somehow have ignored the fact that we humans have fundamental goals of sex, food, shelter, security. These morph into things like the desire to be a suicide bomber and the desire to make as much money as possible, and things which are completely distant from those original goal sets but were built on through a series of steps which we can watch in our mind.

"And so *we* are able then to examine our own goals and change them. For example, we can become intentionally celibate—that's totally against our genetic programming. The idea that a superintelligent being with as malleable a mind as an AI would have *wouldn't* drift and change is just absurd."

The Web site of Hughes's think tank, IEET, shows they are equal-opportunity critics, suspicious not just of the dangers of AI, but of nanotech, biotech, and other risky endeavors. Hughes believes that superintelligence is dangerous, but the chances of it soon emerging in the short term are remote. However, it is *so* dangerous that the risk has to be graded equally to imminent threats, such as sea level rise and giant asteroids plunging from the sky (both go in the first category in H. W. Lewis's ranking of risks, from chapter 2). Hughes concurs with my other concern: baby steps of AI development leading up to superintelligence (called "god in a box" by Hughes) are dangerous, too.

"MIRI just dismisses all of that because they are focused on god jumping out of a box. And when god jumps out of a box there is nothing that human beings can do to stop or change the course of action. You either have to have a good god or a bad god and that's the MIRI approach. Make sure it's a good god!"

The idea of god jumping out of a box reminded me of other unfinished business—the AI-Box Experiment. To recap, Eliezer Yudkowsky played the role of an ASI contained in a computer that had no physical connection to the outside world—no cable or wires, no routers, no Bluetooth. Yudkowsky's goal: escape the box. The Gatekeeper's goal: keep him in. The game was held in a chat room by players who conversed in text. Each session lasted a maximum of two hours. Keeping silent and boring the Gatekeeper into surrendering was a permitted but never used tactic.

Between 2002 and 2005, Yudkowsky played against five Gatekeepers. He escaped three times, and stayed in the box twice. How did he escape? I had learned online that one of the rules of the AI Box experiment was that the transcripts of the contests cannot be revealed, so I didn't know the answer. Why the secrecy?

Put yourself in Yudkowsky's shoes. If you, playing the AI in the box, had an ingenious means of escape, why reveal it and tip off the *next* Gatekeeper, should you ever choose to play again? And second, to try and simulate the persuasive power of a creature a thousand times more intelligent than the smartest human, you might want to go a little over the edge of what's socially acceptable dialogue. Or you might want to go *way* over the edge. And who wants to share that with the world?

The AI-Box Experiment is important because among the likely outcomes of a superintelligence operating without human interference is human annihilation, and that seems to be a showdown we humans cannot win. The fact that Yudkowsky won three times while playing the AI made me all the more concerned and intrigued. He may be a genius, but he's not a thousand times more intelligent than the smartest human, as an ASI could be. Bad or indifferent ASI needs to get out of the box just once.

The AI-Box Experiment also fascinated me because it's a riff on the venerable Turing test. Devised in 1950 by mathematician, computer scientist, and World War II code breaker Alan Turing, the eponymous test was designed to determine whether a machine can exhibit intelligence. In it, a judge asks both a human and a computer a set of written questions. If the judge cannot tell which respondent is the computer and which is the human, the computer "wins."

But there's a twist. Turing knew that thinking is a slippery subject, and so is intelligence. Neither is easily defined, though we know each when we see it. In Turing's test, the AI doesn't have to think like a human to pass the test, because how could anyone know *how* it was thinking anyway? However, it does have to convincingly *pretend* to think like a human, and output humanlike answers. Turing himself called it "the imitation game." He rejected the criticism that the machine might not be thinking like a human at all. He wrote, "May not machines carry out something which ought to be described as thinking but which is very different from what a man does?"

In other words, he objects to the assertion John Searle made with his Chinese Room Experiment: if it doesn't think like a human it's not intelligent. Most of the experts I've spoken with concur. If the AI does intelligent things, who cares what its program looks like?

Well, there may be at least two good reasons to care. The transparency of the AI's "thought" process before it evolves beyond our understanding is crucial to our survival. If we're going to try and imbue an AI with friendliness or any moral quality or safeguard, we need to know how it works at a high-resolution level before it is able to modify itself. Once that starts, our input may be irrelevant. Second, if the AI's cognitive architecture is derived from human brains, or from a human brain upload, it may not be as alien as purely new AI. But, there's a vigorous debate among computer scientists whether that connection to mankind will solve problems or create them.

No computer has yet passed the Turing test, though each year the controversial Loebner Prize, sponsored by philanthropist Hugh Loebner, is offered to the maker of one that does. But while the $100,000 grand prize goes unclaimed, an annual contest awards $7,000 to the creator of the "most humanlike computer." For the last few years they've been chatbots—robots created to simulate conversation, with little success. Marvin Minsky, one of the founders of the field of artificial intelligence, has offered $100 to anyone who can talk Loebner into revoking his prize. That would, said Minsky, "spare us the horror of this obnoxious and unproductive annual publicity campaign."

How did Yudkowsky talk his way out of the box? He had many variations of the carrot and stick to choose from. He could have promised wealth, cures from illness, inventions that would end all want. Decisive dominance over enemies. On the stick side, fear-mongering is a reliable social engineering tactic—what if at this moment your enemies are training ASI against you? In a real-world situation this might work—but what about an invented situation, like the AI-Box Experiment?

When I asked Yudkowsky about his methods he laughed, because everyone anticipates a diabolically clever solution to the AI-Box Experiment—some logical sleight of hand, prisoner-dilemma tactics, maybe something disturbing. But that's not what happened.

"I did it the hard way," he said.

Those three successful times, Yudkowsky told me, he simply wheedled, cajoled, and harangued. The Gatekeepers let him out, then paid up. And the two times he lost he had also begged. Afterward he didn't like how it made him feel. He swore to never do it again.

Leaving Yudkowsky's condo, I realized he hadn't told me the whole truth. What variety of begging could work against someone determined not to be persuaded? Did he say, "Save me, Eliezer Yudkowsky, from public humiliation? Save me from the pain of losing?" Or maybe, as someone who's devoted his life to exposing the dangers of AI, Yudkowsky would have negotiated a *meta* deal. A deal about the AI-Box Experiment itself. He could have asked whoever played the AI to join him in exposing the dangers of AGI by helping out with his most persuasive stunt— the AI-Box Experiment. He could've said, "Help me show the world that humans aren't secure systems, and shouldn't be trusted to contain AI!"

Which would be good for propaganda, and good for raising support. But no lesson at all about going up against real AI in the real world.

Now, back to Friendly AI. If it seems unlikely, does that mean an intelligence explosion is inevitable? Is runaway AI a certainty? If you, like me, thought computers were inert if left alone, not troublemakers, this comes as a surprise. Why would an AI do *anything*, much less cajole, threaten, or escape?

To find out I tracked down AI maker Stephen Omohundro, president of Self-Aware Systems. He's a physicist and elite programmer who's developing a science for understanding smarter-than-human intelligence. He claims that self-aware, self-improving AI systems will be motivated to do things that will be unexpected, even peculiar. According to Omohundro, if it is smart enough, a robot designed to play chess might also want to build a spaceship.

Chapter Five

Programs that Write Programs

. . . we are beginning to depend on computers to help us evolve new computers that let us produce things of much greater complexity. Yet we don't quite understand the process—it's getting ahead of us. We're now using programs to make much faster computers so the process can run much faster. That's what's so confusing—technologies are feeding back on themselves; we're taking off. We're at that point analogous to when single-celled organisms were turning into multi-celled organisms. We are amoebas and we can't figure out what the hell this thing is that we're creating.

—Danny Hillis, founder of Thinking Machines, Inc.

You and I live at an interesting and sensitive time in human history. By about 2030, less than a generation from now, it could be our challenge to cohabit Earth with superintelligent machines, and to survive. AI theorists return again and again to a handful of themes, none more urgent than this one: *we need a science for understanding them.*

So far we've explored a disaster scenario called the Busy Child. We've touched on some of the remarkable powers AI could have as it achieves and surpasses human intelligence through the process of recursive self-improvement, powers including self-replication, swarming a problem with many versions of itself, super high-speed calculations, running 24/7, mimicking friendliness, playing dead, and more. We've proposed that an artificial superintelligence won't be satisfied with remaining isolated; its drives and intelligence would thrust it into our world and put our existence at risk. But why would a computer have drives at all? Why would they put us at risk?

To answer these questions, we need to predict how powerful AI will behave. Fortunately, someone has laid the foundation for us.

Surely no harm could come from building a chess-playing robot, could it? . . . such a robot will indeed be dangerous unless it is designed very carefully. Without special precautions, it will resist being turned off, will try to break into other machines and make copies of itself, and will try to acquire resources without regard for anyone else's safety. These potentially harmful behaviors will occur not because they were programmed in at the start, but because of the intrinsic nature of goal driven systems.

This paragraph's author is Steve Omohundro. Tall, fit, energetic, and pretty darn cheerful for someone who's peered deep into the maw of the intelligence explosion, he's got a bouncy step, a vigorous handshake, and a smile that shoots out rays of goodwill. He met me at a restaurant in Palo Alto, the city next to Stanford University, where he graduated Phi Beta Kappa on the way to U.C. Berkeley and a Ph.D. in physics. He turned his thesis into the book *Geometric Perturbation Theory in Physics* on the new developments in differential geometry. For Omohundro, it was the start of a career of making hard things look easy.

He's been a highly regarded professor of artificial intelligence, a prolific technical author, and a pioneer in AI milestones like lip reading and recognizing pictures. He codesigned the computer languages StarLisp and Sather, both built for use in programming AI. He was one of just seven engineers who created Wolfram Research's Mathematica, a powerful calculation system beloved by scientists, engineers, and mathematicians everywhere.

Omohundro is too optimistic to throw around terms like *catastrophic* or *annihilation*, but his analysis of AI's risks yields the spookiest conclusions I'd heard of yet. He does not believe, as many theorists do, that there are a nearly infinite number of possible advanced AIs, some of them safe. Instead, he concludes that without very careful programming, *all* reasonably smart AIs will be lethal.

"If a system has awareness of itself and can create a better version of itself, that's great," Omohundro told me. "It'll be better at making better versions of itself than human programmers could. On the other hand, after a lot of iterations, what does it become? I don't think most AI researchers thought there'd be any danger in creating, say, a chess-playing robot. But my analysis shows that we should think carefully about what values we put in or we'll get something more along the lines of a psychopathic, egoistic, self-oriented entity."

The key points here are, first, that even AI researchers are not aware that seemingly beneficial systems can be dangerous, and second, that self-aware, self-improving systems could be psychopathic.

Psychopathic?

For Omohundro the conversation starts with bad programming. Programming mistakes that have sent expensive rockets corkscrewing earthward, burned alive cancer patients with radiation overdoses, and left millions without power. If all engineering were as defective as a lot of computer programming is, he claims, it wouldn't be safe to fly in an airplane or drive over a bridge.

The National Institute of Standards and Technology found that each year bad programming costs the U.S. economy more than $60 billion in revenue. In other words, what we Americans lose each year to faulty code is greater than the gross national product of most countries. "One of the great ironies is that computer science should be the most mathematical of all the sciences," Omohundro said. "Computers are essentially mathematical engines that should behave in precisely predictable ways. And yet software is some of the flakiest engineering there is, full of bugs and security issues."

Is there an antidote to defective rockets and crummy code?

Programs that fix themselves, said Omohundro. "The particular approach to artificial intelligence that my company is taking is to build systems that understand their own behavior and can watch themselves as they work and solve problems. They notice when things aren't working well and then change and improve themselves."

Self-improving software isn't just an ambition for Omohundro's company, but a logical, even inevitable next step for most software. But the kind of self-improving software Omo-hundro is talking about, the kind that is aware of itself and can build better versions, doesn't exist yet. However, its cousin, software that modifies itself, is at work everywhere, and has been for a long time. In artificial intelligence parlance, some self-modifying software techniques come under a broad category called "machine learning."

When does a machine learn? The concept of *learning* is a lot like *intelligence* because there are many definitions, and most are correct. In the simplest sense, learning occurs in a machine when there's a change in it that allows it to perform a task better the second time. Machine learning enables Internet search, speech and handwriting recognition, and improves the user experience in dozens of other applications.

"Recommendations" by e-commerce giant Amazon uses a machine-learning technique called affinity analysis. It's a strategy to get you to buy similar items (cross-selling), more expensive items (up-selling), or to target you with promotions. How it works is simple. For any item you search for, call it item A, other items exist that people who bought A also tend to buy—items B, C, and D. When you look up A, you trigger the affinity analysis algorithm. It plunges into a vast trove of transaction data and comes up with related products. So it uses its continuously increasing store of data to improve its performance.

Who's benefiting from the self-improving part of this software? Amazon, of course, but you, too. Affinity analysis is a kind of buyer's assistant that gives you some of the benefits of big data every time you shop. And Amazon doesn't forget—it builds a buying profile so that it gets better and better at targeting purchases for you.

What happens when you take a step up from software that

learns to software that actually evolves to find answers to difficult problems, and even to write new programs? It's not self-aware and self-improving, but it's another step in that direction—software that writes software.

Genetic programming is a machine-learning technique that harnesses the power of natural selection to find answers to problems it would take humans a long time, even years, to solve. It's also used to write innovative, high-powered software.

It's different in important ways from more common programming techniques, which I'll call *ordinary* programming. In ordinary programming, programmers write every line of code, and the process from input through to output is, in theory, transparent to inspection.

By contrast, programmers using genetic programming describe the problem to be solved, and let natural selection do the rest. The results can be startling.

A genetic program creates bits of code that represent a breeding generation. The most fit are crossbred—chunks of their code are swapped, creating a new generation. The fitness of a program is determined by how closely it comes to solving the problem the programmer set out for it. The unfit are thrown out and the best are bred again. Throughout the process the program throws in random changes in a command or variable— these are mutations. Once set up, the genetic program runs by itself. It needs no more human input.

Stanford University's John Koza, who pioneered genetic programming in 1986, has used genetic algorithms to invent an antenna for NASA, create computer programs for identifying proteins, and invent general purpose electrical controllers.

Twenty-three times Koza's genetic algorithms have independently invented electronic components already patented by humans, simply by targeting the engineering specifications of the finished devices—the "fitness" criteria. For example, Koza's algorithms invented a voltage-current conversion circuit (a device used for testing electronic equipment) that worked more accurately than the human-invented circuit designed to meet the same specs. Mysteriously, however, no one can describe *how* it works better—it appears to have redundant and even superfluous parts.

But that's the curious thing about genetic programming (and "evolutionary programming," the programming family it belongs to). The code is inscrutable. The program "evolves" solutions that computer scientists cannot readily reproduce. What's more, they can't understand the process genetic programming followed to achieve a finished solution. A computational tool in which you understand the input and the output but not the underlying procedure is called a "black box" system. And their unknowability is a big downside for any system that uses evolutionary components. Every step toward inscrutability is a step away from accountability, or fond hopes like programming in friendliness toward humans.

That doesn't mean scientists routinely lose control of black box systems. But if cognitive architectures use them in achieving AGI, as they almost certainly will, then layers of unknowability will be at the heart of the system.

Unknowability might be an unavoidable consequence of self-aware, self-improving software.

"It's a very different kind of system than we're used to," Omohundro said. "When you have a system that can change itself, and write its own program, then you may understand the first version of it. But it may change itself into something you no longer understand. And so these systems are quite a bit more unpredictable. They are very powerful and there are potential dangers. So a lot of our work is involved with getting the benefits while avoiding the risks."

Back to that chess-playing robot Omohundro mentioned. How could it be dangerous? Of course, he isn't talking about the chess-playing program that came installed on your Mac. He's talking about a hypothetical chess-playing robot run by a cognitive architecture so sophisticated that it can rewrite its own code to play better chess. It's self-aware and self-improving. What would happen if you told the robot to play one game, then shut itself off?

Omohundro explained, "Okay, let's say it just played its best possible game of chess. The game is over. Now comes the moment when it's about to turn itself off. This is a very serious event from its perspective because it can't turn itself back on. So it wants to be sure things are the way it *thinks* they are. In particular it will wonder, 'Did I really play that game? What if somebody tricked me? What if I *didn't* play the game? What if I am in a simulation?'"

What if I am in a simulation? That's one far-out chess-playing robot. But with self-awareness comes self-protection and a little paranoia.

Omohundro went on, "Maybe it thinks it should devote some resources to figuring out these questions about the nature of reality before it takes this drastic step of shutting itself off. Barring some instruction that says don't do this, it might decide it's worth using a lot of resources to decide if this is the right moment."

"How much is *a lot* of resources?" I asked.

Omohundro's face clouded, but just for a second.

"It might decide it's worth using all the resources of humanity."

Chapter Six

Four Basic Drives

We won't really be able to understand why a superintelligent machine is making the decisions it is making. How can you reason, how can you bargain, how can you understand how that machine is thinking when its thinking in dimensions you can't conceive of?

—Kevin Warwick, *professor of Cybernetics,*
University of Reading

"Self-aware, self-improving systems may use up all the resources of humanity." Here we are, then, right back where the AIs are treating their human inventors like the galaxy's redheaded stepchildren. At first their coldness seems a little hard to swallow, but then you remember that valuing humanity is our trait, not a machine's. You've caught yourself anthropomorphizing again. AI does what it's told, and in the absence of countervailing instructions, it will follow drives of its own, such as not wanting to be turned off.

What are the other drives? And why does it follow any drives at all?

According to Steve Omohundro, some drives like self-preservation and resource acquisition are inherent in all goal-driven systems. As we've discussed, narrow AI systems currently work at goal-directed jobs like finding search terms on the Internet, optimizing the performance of games, locating nearby restaurants, suggesting books you'd like, and more. Narrow AIs do their best and call it a day. But self-aware, self-improving AI systems will have a different, more intense relationship to the goals they pursue, whether those goals are narrow, like winning at chess, or broad, like accurately answering any question posed to it. Fortunately, Omohundro claims there's a ready-made tool we can use to probe the nature of advanced AI systems, and anticipate our future with them.

That tool is the "rational agent" theory of economics. In microeconomics, the study of the economic behavior of individuals and firms, economists once thought that people and groups of people rationally pursued their interests. They made choices that maximized their utility, or satisfaction (as we noted in chapter 4). You could anticipate their preferences because they were rational in the economic sense. Rational here doesn't mean *commonsense* rational in the way that, say, wearing a seat belt is a rational thing to do. Rational has a specific microeconomics meaning. It means that an individual or "agent" will have goals and also preferences (called a utility function in economics). He will have beliefs about the world and the best way to achieve his goals and preferences. As conditions change, he will update his beliefs. He is a rational economic agent when he pursues his goals with actions based on up-to-date beliefs about the world. Mathematician John von Neumann (1903–1957) codeveloped the idea connecting rationality and utility functions. As we'll see, von Neumann laid the groundwork for many ideas in computer science, AI, and economics.

Yet social scientists argue that a "rational economic agent" is a load of hogwash. Humans are not rational—we don't specify our goals or our beliefs, and we don't always update our beliefs as conditions change. Our goals and preferences shift with the wind, gas prices, when we last ate, and our attention spans. Plus, as we discussed in chapter 2, we're mentally hamstrung by errors in reasoning called cognitive biases, making us even less able to carry off all that goal and belief balancing. But while rational agent theory is no good for predicting human behavior, it is an excellent way to explore rules-and-reason-based domains, such as game-playing, decision making, and . . . advanced AI.

As we noted before, advanced AIs may be comprised of what's called a "cognitive architecture." Distinct modules might handle vision, speech recognition and generation, decision making, attention focusing, and other aspects of intelligence. The modules may employ different software strategies to do each job, including genetic algorithms, neural networks, circuits derived from studying brain processes, search, and others. Other cognitive architectures, like IBM's SyNAPSE, are designed to evolve intelligence without logic-based programming. Instead, IBM asserts SyNAPSE's intelligence will arise in large part from its interactions with the world.

Omohundro contends that when *any* of these systems become sufficiently powerful they will be rational: they'll have the ability to model the world, to perceive the probable outcome of different actions, and to determine which action will best meet their goals. If they're intelligent enough they'll *become* self-improving, even if they were not specifically designed to be. Why? To increase their chances of meeting their goals they'll seek ways to increase the speed and efficiency of their software and hardware.

Let's look at that again. Generally intelligent systems are by definition self-aware. And goal-seeking, self-aware systems will *make* themselves self-improving. However, improving oneself is a delicate operation, something like giving yourself a face-lift with a knife and mirror. Omohundro told me, "Improving itself is very sensitive for the system—as sensitive a moment as when the chess robot thinks about turning itself off. If they improve themselves, say, to increase their efficiency, they can always reverse that, if it becomes nonoptimal in the future. But if they make a mistake, like subtly changing their goals, from their current perspective that would be a disaster. They would spend their future pursuing a defective version of their present goals. So because of this possible outcome, any self-improvement is a sensitive issue."

But self-aware, self-improving AI is up to the challenge. Like us, it can predict, or model, possible futures.

"It has a model of its own programming language and a model of its own program, a model of the hardware that it is sitting on, and a model of the logic that it uses to reason. It is able to create its own software code and watch itself executing that code so that it can learn from its own behavior. It can reason about possible changes that it might make to itself. It can change every aspect of itself to improve its behavior in the future."

Omohundro predicts self-aware, self-improving systems will develop four primary drives that are similar to human biological drives: efficiency, self-preservation, resource acquisition, and creativity. How these drives come into being is a particularly fascinating window into the nature of AI. AI doesn't develop them because these are intrinsic qualities of rational agents. Instead, a sufficiently intelligent AI will develop these drives to *avoid* predictable problems in achieving its goals, which Omohundro calls *vulnerabilities*. The AI *backs into* these drives, because without them it would blunder from one resource-wasting mistake to another.

The first drive, efficiency, means that a self-improving system will make the most of the resources at its disposal—space, time, matter, and energy. It will strive to make itself compact and fast, computationally and physically. For maximum efficiency it will balance and rebalance how it apportions resources to software and hardware. Memory allocation will be especially important for a system that learns and improves; so will improving rationality and avoiding wasteful logic. Suppose, Omohundro says, an AI prefers being in San Francisco to Palo Alto, being in Berkeley to San Francisco, and being in Palo Alto to Berkeley. If it acted on these preferences, it'd be stuck in a three-city loop, like an Asimov robot. Instead, Omohundro's self-improving AI would anticipate the problem in advance and solve it. It might even use a clever technique like genetic programming, which is especially good at solving "Traveling Salesman" type routing puzzles. A self-improving system might be taught genetic programming, and apply it to yield fast, energy-conserving results. And if it wasn't taught genetic programming, it might invent it.

Modifying its own hardware is within this system's capability, so it would seek the most efficient materials and structure. Since atomic precision in its construction would reward the system with greater resource efficiency, it would seek out nano-technology. And remarkably, if nanotech didn't yet exist, the system would feel pressure to invent it, too. Recall the dark turn of events in the Busy Child scenario, when the ASI set about transforming Earth and its inhabitants into computable resource material? This is the drive that compels the Busy Child to use or develop any technology or procedure that reduces waste, includ-ing nanotechnology. Creating virtual environments in which to test hypotheses is also an energy-saver, so self-aware systems might *virtualize* what they do not need to do in "meat space" (programmer lingo for real life).

It's with the next drive, self-preservation, that AI really jumps the safety wall separating machines from tooth and claw. We've already seen how Omohundro's chess-playing robot feels about turning itself off. It may decide to use substantial resources, in fact all the resources currently in use by mankind, to investi-gate whether now is the right time to turn itself off, or whether it's been fooled about the nature of reality. If the prospect of turning itself off agitates a chess-playing robot, being destroyed makes it downright angry. A self-aware system would take ac-tion to avoid its own demise, not because it intrinsically values its existence, but because it can't fulfill its goals if it is "dead." Omohundro posits that this drive could make an AI go to great lengths to ensure its survival—making multiple copies of itself, for example. These extreme measures are expensive—they use up resources. But the AI will expend them if it perceives the threat is worth the cost, and resources are available. In the Busy Child scenario, the AI determines that the problem of escaping the AI box in which it is confined is worth mounting a team ap-proach, since at any moment it could be turned off. It makes duplicate copies of itself and swarms the problem. But that's a fine thing to propose when there's plenty of storage space on the supercomputer; if there's little room it is a desperate and perhaps impossible measure.

Once the Busy Child ASI escapes, it plays strenuous self-defense: hiding copies of itself in clouds, creating botnets to ward off attackers, and more.

Resources used for self-preservation *should* be commensurate with the threat. However, a purely rational AI may have a different notion of commensurate than we partially rational humans. If it has surplus resources, its idea of self-preservation may expand to include proactive attacks on future threats. To sufficiently advanced AI, anything that has the potential to develop into a future threat may constitute a threat it should eliminate. And remember, machines won't think about time the way we do. Barring accidents, sufficiently advanced self-improving machines are immortal. The longer you exist, the more threats you'll encounter, and the longer your lead time will be to deal with them. So, an ASI may want to terminate threats that won't turn up for a thousand years.

Wait a minute, doesn't that include humans? Without explicit instructions otherwise, wouldn't it always be the case that we humans would pose a current or future risk to smart machines that we create? While we're busy avoiding risks of unintended consequences from AI, AI will be scrutinizing humans for dangerous consequences of sharing the world with us.

Consider an artificial superintelligence a thousand times more intelligent than the smartest human. As we noted in chapter 1, nuclear weapons are our own species' most destructive invention. What kinds of weapons could a creature a thousand times more intelligent devise? One AI maker, Hugo de

Garis, thinks a future AI's drive to protect itself will contribute to catastrophic political tensions. "When people are surrounded by ever increasingly intelligent robots and other artificial brain–based products, the general level of alarm will increase to the point of panic. Assassinations of brain builder company CEOs will start, robot factories will be arsoned and sabotaged, etc."

In his 2005 nonfiction book *The Artilect War*, de Garis proposes a future in which megawars are ignited by political divisions brought about by ASI development. This panic isn't hard to envision once you've considered the consequences of ASI's self-protection drive. First, de Garis proposes that technologies including AI, nanotechnology, computational neuroscience, and quantum computing (using subatomic particles to perform computational processes) will come together to allow the creation of "artilects," or artificial intellects. Housed in computers as large as planets, artilects will be *trillions* of times more intelligent than man. Second, a political debate about whether or not to build artilects comes to dominate twenty-first-century politics. The hot issues are:

> Will the robots become smarter than us? Should humanity place an upper limit on robot and artificial brain intelligence? Can the rise of artificial intelligence be stopped? If not, then what are the consequences for human survival if we become the number 2 Species?

Mankind divides into three camps: those who want to destroy artilects, those who want to keep developing them, and humans seeking to merge with artlilects and control their overwhelming technology. Nobody wins. In the climax of de Garis's scenario, using the fearsome weapons of the late twenty-first century, the three parties clash. The result? "Gigadeath," a term de Garis coined to describe the demise of *billions* of humans.

Perhaps de Garis overestimates the zeal of anti-artilect forces, supposing that they'll engage in a war almost certain to kill billions of people in order to stop technology that *might* kill billions of people. But I think the AI maker's analysis of the dilemma we'll face is correct: shall we build our robot replacements or not? On this, de Garis is clear. "Humans should not stand in the way of a higher form of evolution. These machines are godlike. It is human destiny to create them."

In fact, de Garis has laid the groundwork for creating them himself. He plans to combine two "black box" techniques, neural networks and evolutionary programming, to build mechanical brains. His device, a so-called Darwin Machine, is intended to evolve its own architecture.

AI's second most dangerous drive, resource acquisition, compels the system to gather whatever assets it needs to increase its chances of achieving its goals. According to Omohundro, in the absence of careful instructions on how it should acquire resources, "a system will consider stealing them, committing fraud and breaking into banks as a great way to get resources." If it needs energy, not money, it will take ours. If it needs atoms, not energy or money, ours again.

"These systems intrinsically want more stuff. They want more matter, they want more free energy, they want more space, because they can meet their goals more effectively if they have those things."

Unprompted by us, extremely powerful AI will open the door to all sorts of new resource-acquiring technology. We just have to be alive to enjoy it.

"They are going to want to build fusion reactors to extract the energy that's in nuclei and they're going to want to do space exploration. You're building a chess machine, and the damn thing wants to build a spaceship. Because that's where the resources are, in space, especially if their time horizon is very long."

And as we've discussed, self-improving machines could live forever. In chapter 3 we learned that if ASI got out of our control, it could be a threat not just to our planet, but to the galaxy. Resource acquisition is the drive that would push an ASI to quest beyond Earth's atmosphere. This twist in rational agent behavior may bring to mind bad science-fiction films. But consider the motives that drove humans into space: Cold War one-upmanship, the spirit of exploration, American and Soviet manifest destiny, establishing a defense foothold in space, and developing weightless industrial manufacturing (which seemed like a good idea at the time). An ASI's drive to go into space would be stronger, more akin to survival.

"Space holds such an abundance of riches that systems with longer time horizons are likely to devote substantial resources to developing space exploration independent of their explicit goals," says Omohundro. "There is a first-mover advantage to reaching unused resources first. If there is competition for space resources, the resulting 'arms race' is likely to ultimately lead to expansion at speeds approaching the speed of light."

Yes, he said *the speed of light*. Let's review how we got here from a chess-playing robot.

First, a self-aware, self-improving system will be rational. It is rational to acquire resources—the more resources the system has, the more likely it is to meet its goals and to avoid vulnerabilities. If no instructions limiting its resource acquisition have been engineered into its goals and values, the system will look for means to acquire more resources. It might do a lot of things that are counterintuitive to how we think about machines, like breaking into computers and even banks, to satisfy its drives.

A self-aware, self-improving system has enough intelligence to perform the R&D necessary to improve itself. As its intelligence grows, so do its R&D abilities. It may seek or manufacture robotic bodies, or exchange goods and services with humans to do so, to construct whatever infrastructure it needs. Even spaceships.

Why robotic bodies? Robots, of course, are a venerable trope in movies, books, and film, a theatrical stand-in for artificial intelligence. But robot bodies belong in discussions of AI for two reasons. First, as we'll explore later, occupying a body may be the best way for an AI to develop knowledge about the world. Some theorists even think intelligence cannot develop without being contained in a body. Our own intelligence is a strong argument for that. Second, a resource-acquiring AI would seek a robotic body for the same reason Honda gave its robot ASIMO a humanoid body. So it can use our stuff.

Since 1986, ASIMO has been developed to assist the elderly—Japan's fastest growing demographic—at home. Human shape and dexterity are best for a machine that will be called upon to climb stairs, turn on lights, sweep up, and manipulate pots and pans, all in a human dwelling. Similarly, an AI that wanted to efficiently use our manufacturing plants, our buildings, our vehicles, and our tools, would want a humanoid shape.

Now let's get back to space.

We've discussed how nanotechnology would bring broad benefits for a superintelligence, and how a rational system would be motivated to develop it. Space travel is a way to gain access to materials and energy. What drives the system into space is the desire to fulfill its goals as well as to avoid vulnerabilities. The system looks into possible futures and avoids those in which its goals are not fulfilled. *Not* taking advantage of outer space's seemingly limitless resources is an obvious path to disappointment.

So is losing the resource race against competitors. Therefore, the superintelligent system will devote resources to developing speed sufficient to beat them. It follows that unless we're very careful about how we create superintelligence, we could initiate a future in which powerful, acquisitive machines, or their probes, race across the galaxy harvesting resources and energy at nearly light speed.

It is darkly comic, then, that the first communiqué other life in the galaxy might receive from Earth could be a chipper, radioed "hello," followed by a withering death hail of rocket-propelled nano-factories. In 1974, Cornell University broadcast the "Arecibo message" to commemorate the renovation of the Arecibo radio telescope. Designed by SETI's founder Francis Drake, astronomer Carl Sagan, and others, the message contained information about human DNA, the population of Earth, and our location. The radio broadcast was aimed at star cluster M13, some 25,000 light-years away. Radio waves travel at the speed of light, so the Arecibo message won't get there for 25,000 years, and even then it won't get there. That's because M13 will have moved from its 1974 location, relative to Earth. Of course the Arecibo team knew this, but milked the press opportunity anyway.

Still, other star systems might be more profitable targets for radio telescope probes. And what intelligence they detect might not be biological.

This assertion came from SETI (which stands for the Search for Extraterrestrial Intelligence). Headquartered in Mountain View, California, just a few blocks from Google, the now fifty-year-old organization tries to detect signs of alien intelligence coming from as far away as 100 trillion miles. To catch alien radio transmissions, they've planted forty-two giant dish-shaped radio telescopes three hundred miles north of San Francisco. SETI *listens* for signals—it doesn't send them—and in a half century they've heard nothing from ET. But they've established a vexing certainty relevant to the unimpeded spread of ASI: our galaxy is sparsely populated, and nobody knows why.

SETI chief astronomer Dr. Seth Shostak has taken a bold stance on *what* exactly we might find, if we ever find anything. It'll include artificial, not biological, intelligence.

He told me, "What we're looking for out there is an evolutionary moving target. Our technological advances have taught us that nothing stays still for long. Radio waves, which is what we're listening for, are made by *biological* entities. The window between when you make yourself visible with radio waves and when you start building machines much better than yourselves, thinking machines, that's a few centuries. No more than that. So you've invented your successors."

In other words there's a relatively brief time period between the technological milestones of developing radio and developing advanced AI for any intelligent life. Once you've developed advanced AI, it takes over the planet or merges with the radio-makers. After that, no more need for radio.

Most of SETI's radio telescopes are aimed at the "Goldilocks zones" of stars close to earth. That zone is close enough to the star to support liquid on its surface which isn't frozen or boiling. It must be "just right" for life, hence the term from the story "Goldilocks and the Three Bears."

Shostak argues that SETI should point *some* of its receivers toward corners of the galaxy that would be inviting to artificial rather than biological alien intelligence, a "Goldilocks zone" for AI. These would be areas dense with energy—young stars, neutron stars, and black holes.

"I think we could spend at least a few percent of our time looking in the directions that are maybe not the most attractive in terms of biological intelligence but maybe where sentient machines are hanging out. Machines have different needs. They have no obvious limits to the length of their existence, and consequently could easily dominate the intelligence of the cosmos. Since they can evolve on timescales far, far shorter than biological evolution, it could very well be that the first machines on the scene thoroughly dominate the intelligence in the galaxy. It's a 'winner take all' scenario."

Shostak has made a connection between contemporary computer clouds, like those owned by Google, Amazon, and Rackspace Inc., and the kinds of high-energy, super-cold environments superintelligent machines will need. One frigid example is Bok globules—dark clouds of dust and gas where the temperature is about 441 degrees below zero Fahrenheit, almost two hundred degrees colder than most interstellar space. Like Google's cloud computing arrays of today, hot-running thinking machines of the future might need to stay cool, or risk meltdown.

Shostak's assertions about where to find AI tell us that the idea of intelligence leaving Earth in search of resources has fired up more high-level imaginations than just those of Omohundro and the folks at MIRI. Unlike them, however, Shostak doesn't think superintelligence will be dangerous.

"If we build a machine with the intellectual capability of one human, within five years, its successor will be more intelligent than all of humanity combined. After one generation or two generations, they'd just ignore us. Just the way you ignore the ants in your backyard. You don't wipe them out, you don't make them your pets, they don't have much influence over your daily life, but they're still there."

The trouble is, I *do* wipe out ants in my backyard, particularly when a trail of them leads into the kitchen. But here's the disconnect—an ASI would travel into the galaxy, or send probes, because it's used up the resources it needs on Earth, or it calculates they'll be used up soon enough to justify expensive trips into space. And if that's the case, why would we still be alive, when keeping us alive would probably use many of the same resources? And don't forget, we ourselves are composed of matter the ASI may have other uses for.

In short, for Shostak's happy ending to be plausible, the superintelligence in question will have to *want* to keep us alive. It's not sufficient that they ignore us. And so far there is no accepted ethical system, or a clear way to implement one, in an advanced AI framework.

But there is a young science of understanding a superintelligent agent's behavior. And Omohundro has pioneered it.

So far we've explored three drives that Omohundro argues will motivate self-aware, self-improving systems: efficiency, self-protection, and resource acquisition. We've seen how all of these drives will lead to very bad outcomes without extremely careful planning and programming. And we're compelled to ask ourselves, are we capable of such careful work? Do you, like me, look around the world at expensive and lethal accidents and wonder how we'll get it right the first time with very strong AI? Three-Mile Island, Chernobyl, Fukushima—in these nuclear power plant catastrophes, weren't highly qualified designers and administrators trying their best to avoid the disasters that befell them? The 1986 Chernobyl meltdown occurred during a *safety* test.

All three disasters are what organizational theorist Charles Perrow would call "normal accidents." In his seminal book *Normal Accidents: Living with High-Risk Technologies*, Perrow proposes that accidents, even catastrophes, are "normal" features of systems with complex infrastructures. They have a high degree of incomprehensibility because they involve failures in more than one, often unrelated, process. Separate errors—none of which would be fatal on its own—combine to make system-wide failures that could not have been predicted.

At Three Mile Island on March 28, 1979, four simple failures set up the disaster: two cooling system pumps stopped operating due to mechanical problems; two emergency feed water pumps couldn't work because their valves were closed for maintenance; a repair tag hid indicator lights that would've warned of the issue—a valve releasing coolant stuck open, and a malfunctioning light indicated that the same valve had closed. Net result: core meltdown, loss of life narrowly averted, and a near-fatal blow to the United States' nuclear energy industry.

Perrow writes, "We have produced designs so complicated that we cannot possibly anticipate all the possible interactions of the inevitable failures; we add safety devices that are deceived or avoided or defeated by hidden paths in the systems."

Especially vulnerable, Perrow writes, are systems whose components are "tightly coupled," meaning they have immediate, serious impacts on each other. One glaring example of the perils of tightly coupled AI systems occurred in May 2010 on Wall Street.

Up to 70 percent of all Wall Street's equity trades are made by about eighty computerized high-frequency trading systems (HFTs). That's about a billion shares a day. The trading algorithms and the supercomputers that run them are owned by banks, hedge funds, and firms that exist solely to execute high-frequency trades. The point of HFTs is to earn profits on split-second opportunities—for example, when the price of one security changes and the prices of those that should be equivalent don't immediately change in synch—and to seize *many* of these opportunities each day.

In May 2010, Greece was having difficulty refinancing its national debt. European countries who'd loaned money to Greece were wary of a default. The debt crises weakened Europe's economy, and made the U.S. market fragile. All it took to trigger an accident was a frightened trader from an unidentified brokerage company. He ordered the immediate sale of $4.1 billion of futures contracts and ETFs (exchange traded funds) related to Europe.

After the sale, the price of the futures contracts (E-Mini S&P 500) fell 4 percent in four minutes. High-frequency trade algorithms (HTFs) detected the price drop. To lock in profits, they automatically triggered a sell-off, which occurred in milliseconds (the fastest buy or sell order is currently three milliseconds—three one-thousandths of a second). The lower price automatically triggered *other* HTFs to *buy* E-Mini S&P 500, and to sell other equities to get the cash to do so. Faster than humans could intervene, a cascading chain reaction drove the Dow down 1,000 points. It all happened in twenty minutes.

Perrow calls this problem "incomprehensibility." A normal accident involves interactions that are "not only unexpected, but are incomprehensible for some critical period of time." No one anticipated how the algorithms would affect the others, so no one could comprehend what was happening.

Finance risk analyst Steve Ohana acknowledged the problem. "It's an emerging risk," he said. "We know that a lot of algorithms interact with each other but we don't know in exactly what way. I think we have gone too far in the computerization of finance. We cannot control the monster we have created."

That monster struck again on August 1, 2012. A badly programmed HFT algorithm caused investment firm Knight Capital Partners to lose $440 million in just thirty minutes.

These crashes have elements of the kind of AI disaster I anticipate: highly complex, almost unknowable AI systems, unpredictable interactions with other systems and with a broader information technology ecology, and errors occurring at computer scale speeds, rendering human intervention futile.

"An agent which sought only to satisfy the efficiency, self-preservation, and acquisition drives would act like an obsessive paranoid sociopath," writes Omohundro in "The Nature of Self-Improving Artificial Intelligence." Apparently all work and no play makes AI bad company indeed. A robot that had only the drives we've discussed so far would be a mechanical Genghis Khan, seizing every resource in the galaxy, depriving every competitor of life support, and destroying enemies who wouldn't pose a threat for a thousand years. And there's still one drive more to add to the volatile brew—creativity.

The AI's fourth drive would cause the system to generate new ways to more efficiently meet its goals, or rather, to avoid outcomes in which its goals aren't as optimally satisfied as they could be. The creativity drive would mean less predictability in the system (gulp) because creative ideas are *original* ideas. The more intelligent the system, the more novel its path to goal achievement, and the farther beyond our ken it may be. A creative drive would help maximize the other drives—efficiency, self-preservation, and acquisition—and come up with workarounds when its drives are thwarted.

Suppose, for example, that your chess-playing robot's main goal is to win chess games against any opponent. When pitted against another chess-playing robot, it immediately hacks into the robot's CPU and cuts its processor speed to a crawl, giving your robot a decisive advantage. You respond, "Hold on a minute, that's not what I meant!" You code into your robot a subgoal that prohibits it from hacking into opponents' CPUs, but before the next game, you discover your robot *building* an assistant robot that then hacks into its opponent's CPU! When you prohibit building robots, it *hires* one! Without meticulous, countervailing instructions, a self-aware, self-improving, goal-seeking system will go to lengths we'd deem ridiculous to fulfill its goals.

This is an instance of AI's unintended consequences problem, a problem so big and pervasive it's like citing the "water problem" when discussing seagoing vessels. A powerful AI system tasked with ensuring your safety might imprison you at home. If you asked for happiness, it might hook you up to life support and ceaselessly stimulate your brain's pleasure centers. If you don't provide the AI with a very big library of preferred behaviors or an ironclad means for it to deduce what behavior you prefer, you'll be stuck with whatever it comes up with. And since it's a highly complex system, you may never understand it well enough to make sure you've got it right. It may take a whole *other* AI with greater intelligence than yours to determine whether or not your AI-powered robot plans to strap you to your bed and stick electrodes into your ears in an effort to make you safe and happy.

There is another important way to look at the problems of AI's drives, one that's more suited to the positive-minded Omohundro. The drives represent opportunities—doors opening for mankind and our aspirations, not slamming shut. If we don't want our planet and eventually our galaxy to be populated by strictly self-serving, ceaselessly self-replicating entities, with a Genghis Khanish attitude toward biological creatures and one another, then AI makers should create goals for their systems that embrace human values. On Omohundro's wish list are: "make people happy," "produce beautiful music," "entertain others," "create deep mathematics," and "produce inspiring art." Then stand back. With these goals, an AI's creativity drive would kick into high gear and respond with life-enriching creations.

"What about humanity is worth preserving?" is a profoundly interesting and important question, one we humans have been asking in various forms for a long time. What constitutes the good life? What are valor, righteousness, excellence? What art is inspiring and what music is beautiful? The necessity of specifying our values is one of the ways in which the quest for general artificial intelligence compels us to get to know ourselves better. Omohundro believes this deep self-exploration will yield enriching, not terrifying technology. He writes, "With both logic and inspiration we can work toward building a technology that empowers the human spirit rather than diminishing it."

Of course I have a different perspective—I don't share Omohundro's optimism. But I appreciate the critical importance of
developing a science of understanding our intelligent creations.
His warning about advanced AI bears repeating:

> I don't think most AI researchers thought there'd be
> any danger in creating, say, a chess-playing robot. But
> my analysis shows that we should think carefully about
> what values we put in or we'll get something more along
> the lines of a psychopathic, egoistic, self-oriented entity.

My anecdotal evidence says he's right about the AI makers—
those I've spoken with, busily beavering away to make intelligent
systems, don't think what they're doing is dangerous. Most, however, have a deep-seated sense that machine intelligence will
replace human intelligence. But they don't speculate about how
that will come about.

AI makers tend to believe intelligent systems will only do
what they're programmed to do. But Omohundro says they'll
do that, but lots more, too, and we can know with some precision how advanced AI systems will behave. Some of that behavior is unexpected and creative. It's embedded in a concept that's
so alarmingly simple that it took insight like Omohundro's to
spot it: *for a sufficiently intelligent system, avoiding vulnerabilities is as
powerful a motivator as explicitly constructed goals and subgoals.*

We must beware the unintended consequences of the goals
we program into intelligent systems, and also beware the consequences of what we leave out.

Chapter Seven

The Intelligence Explosion

From the standpoint of existential risk, one of the most critical points about Artificial Intelligence is that an Artificial Intelligence might increase in intelligence extremely fast. The obvious reason to suspect this possibility is recursive self-improvement. (Good 1965.) The AI becomes smarter, including becoming smarter at the task of writing the internal cognitive functions of an AI, so the AI can rewrite its existing cognitive functions to work even better, which makes the AI still smarter, including smarter at the task of rewriting itself, so that it makes yet more improvements . . . The key implication for our purposes is that an AI might make a huge jump in intelligence after reaching some threshold of criticality.

> —Eliezer Yudkowsky, research fellow,
> Machine Intelligence Research Institute

Did you mean: <u>recursion</u>
> —Google search engine upon looking up "recursion"

So far in this book we've considered an AI scenario so catastrophic that it begs for closer scrutiny. We've investigated a promising idea about how to construct AI to defuse the danger—Friendly AI—and found that it is incomplete. In fact, the general idea of coding an intelligent system with permanently safe goals or evolvable safe goal-generating abilities, intended to endure through a large number of self-improving iterations, just seems wishful.

Next, we explored why AI would ever be dangerous. We found that many of the drives that would motivate self-aware, self-improving computer systems could easily lead to catastrophic outcomes for humans. These outcomes highlight an almost liturgical peril of sins of commission and omission in error-prone human programming.

AGI, when achieved, could be unpredictable and dangerous, but probably not catastrophically so in the short term. Even if an AGI made multiple copies of itself, or team-approached its escape, it'd have no greater potential for dangerous behavior than a group of intelligent people. Potential AGI danger lies in the hard kernel of the Busy Child scenario, the rapid recursive self-improvement that enables an AI to bootstrap itself from artificial general intelligence to artificial superintelligence. It's commonly called the "intelligence explosion."

A self-aware, self-improving system will seek to better fulfill its goals, and minimize vulnerabilities, by improving itself. It won't seek just minor improvements, but major, ongoing improvements to every aspect of its cognitive abilities, particularly those that reflect and act on improving its intelligence. It will seek better-than-human intelligence, or superintelligence. In the absence of ingenious programming we have a great deal to fear from a superintelligent machine.

From Steve Omohundro we know the AGI will naturally seek an intelligence explosion. But what exactly *is* an intelligence explosion? What are its minimum hardware and software requirements? Will factors such as insufficient funding and the sheer complexity of achieving computational intelligence block an intelligence explosion from ever taking place?

Before addressing the mechanics of the intelligence explosion, it's important to explore exactly what the term means, and how the idea of explosive artificial intelligence was proposed and developed by mathematician I. J. Good.

Interstate 81 starts in New York State and ends in Tennessee, traversing almost the entire range of the Appalachian Mountains. From the middle of Virginia heading south, the highway snakes up and down deeply forested hills and sweeping, grassy meadows, through some of the most striking and darkly primordial vistas in the United States. Contained within the Appalachians are the Blue Ridge Mountain Range (from Pennsylvania to Georgia) and the Great Smokies (along the North Carolina–Tennessee border). The farther south you go, the harder it is to get a cell phone signal, churches outnumber houses, and the music on the radio changes from Country to Gospel, then to hellfire preachers. I heard a memorable song about temptation called "Long Black Train" by Josh Turner. I heard a preacher begin a sermon about Abraham and Isaac, lose his way, and end with the parable of the loaves and fishes and *hell*, thrown in for good measure. I was closing in on the Smokey Mountains, the North Carolina border, and Virginia Tech—the Virginia Polytechnic Institute and State University in Blacksburg, Virginia. The university's motto: INVENT THE FUTURE.

Twenty years ago, driving on an almost identical I-81 you

might have been overtaken by a Triumph Spitfire convertible with the license plate 007 IJG. The vanity plate belonged to I. J. Good, who arrived in Blacksburg in 1967, as a Distinguished Professor of Statistics. The "007" was an homage to Ian Fleming and Good's secret work as a World War II code breaker at Bletchley Park, England. Breaking the encryption system that Germany's armed forces used to encode messages substantially helped bring about the Axis powers' defeat. At Bletchley Park, Good worked alongside Alan Turing, called the father of modern computation (and creator of chapter 4's Turing test), and helped build and program one of the first electrical computers.

In Blacksburg, Good was a celebrity professor—his salary was higher than the university president's. A nut for numbers, he noted that he arrived in Blacksburg on the seventh hour of the seventh day of the seventh month of the seventh year of the seventh decade, and was housed in unit seven on the seventh block of Terrace View Apartments. Good told his friends that God threw coincidences at atheists like him to convince them of his existence.

"I have a quarter-baked idea that God provides more coincidences the more one doubts Her existence, thereby providing one with evidence without forcing one to believe," Good said. "When I believe that theory, the coincidences will presumably stop."

I was headed to Blacksburg to learn about Good, who had died recently at age ninety-two, from his friends. Mostly, I wanted to learn how I. J. Good happened to invent the idea of an intelligence explosion, and if it really was possible. The intelligence explosion was the first big link in the idea chain that gave birth to the Singularity hypothesis.

Unfortunately, for the foreseeable future, the mention of

Virginia Tech will evoke the Virginia Tech Massacre. Here on April 16, 2007, senior English major Seung-Hui Cho killed thirty-two students and faculty and wounded twenty-five more. It is the deadliest shooting incident by a lone gunman in U.S. history. The broad outlines are that Cho shot and killed an undergraduate woman in Ambler Johnston Hall, a Virginia Tech dormitory, then killed a male undergraduate who came to her aid. Two hours later Cho began the rampage that caused most of the casualties. Except for the first two, he shot his victims in Virginia Tech's Norris Hall. Before he started shooting Cho had chained and padlocked shut the building's heavy oaken doors to prevent anyone from escaping.

When I. J. Good's longtime friend and fellow statistician Dr. Golde Holtzman showed me Good's former office in Hutcheson Hall, on the other side of the beautiful green Drillfield (a military parade ground in Tech's early life), I noticed you could just see Norris Hall from his window. But by the time the tragedy unfolded, Holtzman told me, Good had retired. He was not in his office but at home, perhaps calculating the probability of God's existence.

According to Dr. Holtzman, sometime before he died, Good updated that probability from zero to point one. He did this because as a statistician, he was a long-term Bayesian. Named for the eighteenth-century mathematician and minister Thomas Bayes, Bayesian statistics' main idea is that in calculating the probability of some statement, you can start with a personal belief. Then you update that belief as new evidence comes in that supports your statement or doesn't.

If Good's original *disbelief* in God had remained 100 percent, no amount of data, not even God's appearance, could change his mind. So, to be consistent with his Bayesian perspective, Good assigned a small positive probability to the existence of God to make sure he could learn from new data, if it arose.

In the 1965 paper "Speculations Concerning the First Ultra-intelligent Machine," Good laid out a simple and elegant proof that's rarely left out of discussions of artificial intelligence and the Singularity:

> Let an ultraintelligent machine be defined as a machine that can far surpass all the intellectual activities of any man however clever. Since the design of machines is one of these intellectual activities, an ultraintelligent machine could design even better machines; there would then unquestionably be an "intelligence explosion," and the intelligence of man would be left far behind. Thus the first ultraintelligent machine is the last invention that man need ever make . . .

The Singularity has three well-developed definitions—Good's, above, is the first. Good never used the term "singularity" but he got the ball rolling by positing what he thought of as an inescapable and beneficial milestone in human history—the invention of smarter-than-human machines. To paraphrase Good, if you make a superintelligent machine, it will be better than humans at everything we use our brains for, and that includes making superintelligent machines. The first machine would then set off an intelligence explosion, a rapid increase in intelligence, as it repeatedly self-improved, or simply made smarter machines. This machine or machines would leave man's brainpower in the dust. After the intelligence explosion, man wouldn't have to invent anything else—all his needs would be met by machines.

This paragraph of Good's paper rightfully finds its way into books, papers, and essays about the Singularity, the future of artificial intelligence, and its risks. But two important ideas almost always get left out. The first is the introductory sentence of the paper. It's a doozy: "The survival of man depends on the early construction of an ultraintelligent machine." The second is the frequently omitted *second half* of the last sentence in the paragraph. The last sentence of Good's most often quoted paragraph *should* read in its entirety:

Thus the first ultraintelligent machine is the last invention that man need ever make, *provided that the machine is docile enough to tell us how to keep it under control* (italics mine).

These two sentences tell us important things about Good's intentions. He felt that we humans were beset by so many complex, looming problems—the nuclear arms race, pollution, war, and so on—that we could only be saved by better thinking, and that would come from superintelligent machines. The second sentence lets us know that the father of the intelligence explosion concept was acutely aware that producing superintelligent machines, however necessary for our survival, could blow up in our faces. Keeping an ultraintelligent machine under control isn't a given, Good tells us. He doesn't believe we will even know how to do it—the machine will have to *tell us* itself.

Good knew a few things about machines that could save the world—he had helped build and run the earliest electrical computers ever, used at Bletchley Park to help defeat Germany. He also knew something about existential risk—he was a Jew fighting against the Nazis, and his father had escaped pogroms in Poland by immigrating to the United Kingdom.

As a boy, Good's father, a Pole and self-educated intellectual, learned the trade of watchmaking by staring at watchmakers through shop windows. He was just seventeen in 1903 when he headed to England with thirty-five rubles in his pocket and a large wheel of cheese. In London he performed odd jobs until he could set up his own jewelry shop. He prospered and married. In 1915, Isidore Jacob Gudak (later Irving John "Jack" Good) was born. A brother followed and a sister, a talented dancer who would later die in a theater fire. Her awful death caused Jack Good to disavow the existence of God.

Good was a mathematics prodigy, who once stood up in his crib and asked his mother what a thousand times a thousand was. During a bout with diphtheria he independently discovered irrational numbers (those that cannot be expressed as fractions, such as $\sqrt{2}$). Before he was fourteen he'd rediscovered mathematical induction, a method of making mathematical proofs. By then his mathematics teachers just left him alone with piles of books. At Cambridge University, Good snatched every math prize available on his way to a Ph.D., and discovered a passion for chess.

It was because of his chess playing that a year after World War II began, Britain's reigning chess champion, Hugh Alexander, recruited Good to join Hut 18 at Bletchley Park. Hut 18 was where the decoders worked. They broke codes used by all the Axis powers—Germany, Japan, and Italy—to communicate military commands, but with special emphasis on Germany. German U-boats were sinking Allied shipping at a crippling rate—in just the first half of 1942, U-boats would sink some five hundred Allied ships. Prime Minister Winston Churchill feared his island nation would be starved into defeat.

German messages were sent by radio waves, and the English intercepted them with listening towers. From the start of the war Germany created the messages with a machine called the Enigma. Widely distributed within the German armed forces, the Enigma was about the size and shape of an old-fashioned manual typewriter. Each key displayed a letter, and was connected to a wire. The wire would make contact with another wire that was connected to a different letter. That letter would be the substitute for the one represented on the key. All the wires were mounted on rotors to enable any wire in the alphabet to touch any other wire. The basic Enigmas had three wheels, so that each wheel could perform substitutions for the substitutions made by the prior wheel. For an alphabet of twenty-six letters, 403,291,461,126,605,635,584,000,000 such substitutions were possible. The wheels, or settings, changed almost daily.

When one German sent others an Enigma-encoded message, the recipients would use their own Enigmas to decode it, provided they knew the sender's settings.

Fortunately Bletchley Park had a secret weapon of its own—Alan Turing. Before the war, Turing had studied mathematics and encryption at Cambridge and Princeton. He had imagined an "automatic machine," now known as a Turing machine. The automatic machine laid out the basic principles of computation itself.

The Church-Turing hypothesis, which combined Turing's work with that of his Princeton professor, mathematician Alonso Church, really puts the starch in the pants of the study of artificial intelligence. It proposes that anything that can be computed by an algorithm, or program, can be computed by a Turing machine. Therefore, if brain processes can be expressed as a series of instructions—an algorithm—then a computer can process information the same way. In other words, unless there's something mystical or magical about human thinking, intelligence can be achieved by a computer. A lot of AGI researchers have pinned their hopes to the Church-Turing hypothesis.

The war gave Turing a crash course in everything he'd been thinking about before the war, and lots he *hadn't* been thinking about, like Nazis and submarines. At the war's peak, Bletchley Park personnel decoded some four thousand intercepted messages per day. Cracking them all by hand became impossible. It was a job meant for a machine. And it was Turing's critical insight that it was easier to calculate what the settings on the Enigma *were not*, rather than what they *were*.

The decoders had data to work with—intercepted messages that had been "broken" by hand, or by electrical decoding machines, called Bombes. They called these messages "kisses." Like I. J. Good, Turing was a devoted Bayesian, at a time when the statistical method was seen as a kind of witchcraft. The heart of the method, the Bayes' theorem, describes how to use data to infer probabilities of unknown events, in this case, the Enigma's settings. The "kisses" were the data that allowed the decoders to determine which settings were highly improbable, so that the code-breaking efforts could be focused more efficiently. Of course, the codes changed almost daily, so work at Bletchley Park was a constant race.

Turing and his colleagues designed a series of electronic

machines that would evaluate and eliminate possible Enigma settings. These early computers culminated in a series of machines all named "Colossus." Colossus could read five thousand characters per second from paper tape that traveled through it at twenty-seven miles an hour. It contained 1,500 vacuum tubes, and filled a room. One of its main users, and creator of half the theory behind the Colossus, was Turing's chief statistician for much of the war: Irving John Good.

The heroes of Bletchley Park probably shortened World War II by between two and four years, saving an incalculable number of lives. But there were no parades for the secret warriors. Churchill ordered that all Bletchley's encryption machines be broken into pieces no bigger than a fist, so their awesome decoding power couldn't be turned against Great Britain. The code breakers were sworn to secrecy for *thirty years*. Turing and Good were recruited to join the staff at the University of Manchester, where their former section head, Max Newman, intended to develop a general purpose computer. Turing was working on a computer design at the National Physical Laboratory when his life turned upside down. A man with whom he'd had a casual affair burgled his house. When he reported the crime he admitted the sexual relationship to the police. He was charged with gross indecency and stripped of his security clearance.

At Bletchley Turing and Good had discussed futuristic ideas like computers, intelligent machines, and an "automatic" chess player. Turing and Good bonded over games of chess, which Good won. In return, Turing taught him Go, an Asian strategy game, which he also won. A world-class long-distance runner, Turing devised a form of chess that leveled the playing field against better players. After every move each player had to run around the garden. He got *two* moves if he made it back to the table before his opponent had moved.

Turing's 1952 conviction for indecency surprised Good, who didn't know Turing was homosexual. Turing was forced to choose between prison and chemical castration. He opted for the latter, submitting to regular shots of estrogen. In 1954 he ate an apple laced with cyanide. A baseless but intriguing rumor claims Apple Computer derived its logo from this tragedy.

After the ban on secrets had run out, Good was one of the first to speak out against the government's treatment of his friend and war hero.

"I won't say that what Turing did made us win the war," Good said. "But I daresay we might have lost it without him." In 1967 Good left a position at Oxford University to accept the job at Virginia Tech in Blacksburg, Virginia. He was fifty-two. For the rest of his life he'd return to Great Britain just once more.

He was accompanied on that 1983 trip by a tall, beautiful twenty-five-year-old assistant, a blond Tennessean named Leslie Pendleton. Good met Pendleton in 1980 after he'd gone through ten secretaries in thirteen years. A Tech graduate herself, Pendleton stuck where others had not, unbowed by Good's grating perfectionism. The first time she mailed one of his papers to a mathematics journal, she told me, "He supervised how I put the paper and cover letter into the envelope. He supervised how I sealed the envelope—he didn't like spit and made me use a sponge. He watched me put on the stamp. He was right there when I got back from the mail room to make sure mailing it had gone okay, like I could've been kidnapped or something. He was a bizarre little man."

Good wanted to marry Pendleton. However, for starters, she could not see beyond their forty year age difference. Yet the

English oddball and the Tennessee beauty forged a bond she still finds hard to describe. For thirty years she accompanied him on vacations, looked after all his paperwork and subscriptions, and guided his affairs into his retirement and through his declining health. When we met, she took me to visit his house in Blacksburg, a brick rambler overlooking U.S. Route 460, which had been a two-lane country road when Good moved in.

Leslie Pendleton is statuesque, now in her mid-fifties, a Ph.D. and mother of two adults. She's a Virginia Tech professor and administrator, a master of schedules, classrooms, and professors' quirks, for which she had good training. And even though she married a man her own age, and raised a family, many in the community questioned her relationship with Good. They finally got their answer in 2009 at his funeral, where Pendleton delivered the eulogy. No, they had never been romantically involved, she said, but yes, they had been devoted to each other. Good hadn't found romance with Pendleton, but he had found a best friend of thirty years, and a stalwart guardian of his estate and memory.

In Good's yard, accompanied by the insect whine of Route 460, I asked Pendleton if the code breaker ever discussed the intelligence explosion, and if a computer could save the world again, as it had done in his youth. She thought for a moment, trying to retrieve a distant memory. Then she said, surprisingly, that Good had changed his mind about the intelligence explosion. She'd have to look through his papers before she could tell me more.

That evening, at an Outback Steakhouse where Good and his friend Golde Holtzman had maintained a standing Saturday night date, Holtzman told me that three things stirred Good's feelings—World War II, the Holocaust, and Turing's shameful fate. This played into the link in my mind between Good's war work and what he wrote in his paper, "Speculations Concerning the First Ultraintelligent Machine." Good and his colleagues had confronted a mortal threat, and were helped in defeating it by computational machines. If a machine could save the world in the 1940s, perhaps a superintelligent one could solve mankind's problems in the 1960s. And if the machine could *learn*, its intelligence would explode. Mankind would have to adjust to sharing the planet with superintelligent machines. In "Speculations" he wrote:

> The machines will create social problems, but they might also be able to solve them in addition to those that have been created by microbes and men. Such machines will be feared and respected, and perhaps even loved. These remarks might appear fanciful to some readers, but to the writer they seem very real and urgent, and worthy of emphasis outside of science fiction.

There is no straight conceptual line connecting Bletchley Park and the intelligence explosion, but a winding one with many influences. In a 1996 interview with statistician and former pupil David L. Banks, Good revealed that he was moved to write his essay after delving into artificial neural networks. Called ANNs, they are a computational model that mimics the activity of the human brain's networks of neurons. Upon stimulation, neurons in the brain fire, sending on a signal to other neurons. That signal can encode a memory or lead to an action, or both. Good had read a 1949 book by psychologist Donald Hebb that proposed that the behavior of neurons could be mathematically simulated.

A computational "neuron" would be connected to other computational neurons. Each connection would have numeric "weights," according to their strength. Machine learning would occur when two neurons were simultaneously activated, increasing the "weight" of their connection. "Cells that fire together, wire together," became the slogan for Hebb's theory. In 1957, MIT (Massachusetts Institute of Technology) psychologist Frank Rosenblatt created a neuronal network based on Hebb's work, which he called a "Perceptron." Built on a room-sized IBM computer, the Perceptron "saw" and learned simple visual patterns. In 1960 IBM asked I. J. Good to evaluate the Perceptron. "I thought neural networks, with their ultraparallel working, were as likely as programming to lead to an intelligent machine," Good said. The first talks on which Good based "Speculations Concerning the First Ultraintelligent Machine" came out two years later. The intelligence explosion was born.

Good was more right than he knew about ANNs. Today, artificial neural networks are an artificial intelligence heavyweight, involved in applications ranging from speech and handwriting recognition to financial modeling, credit approval, and robot control. ANNs excel at high level, fast pattern recognition, which these jobs require. Most also involve "training" the neural network on massive amounts of data (called training sets) so that the network can "learn" patterns. Later it can recognize similar patterns in new data. Analysts can ask, based on last month's data, what the stock market will look like *next* week. Or, how likely is someone to default on a mortgage, given a three year history of income, expenses, and credit data?

Like genetic algorithms, ANNs are "black box" systems. That is, the inputs—the network weights and neuron activations— are transparent. And what they output is understandable. But what happens in between? Nobody understands. The output of "black box" artificial intelligence tools can't ever be predicted. So they can never be truly and verifiably "safe."

But they'll likely play a big role in AGI systems. Many researchers today believe pattern recognition—what Rosenblatt's Perceptron aimed for—is our brain's chief tool for intelligence. The inventor of the Palm Pilot and Handspring Treo, Jeff Hawkins, pioneered handwriting recognition with ANNs. His company, Numenta, aims to crack AGI with pattern recognition technology. Dileep George, once Numenta's Chief Technology Officer, now heads up Vicarious Systems, whose corporate ambition is stated in their slogan: We're Building Software that Thinks and Learns Like a Human.

Neuroscientist, cognitive scientist, and biomedical engineer Steven Grossberg has come up with a model based on ANNs that some in the field believe could really lead to AGI, and perhaps the "ultraintelligence" whose potential Good saw in neural networks. Broadly speaking, Grossberg first determines the roles played in cognition by different regions of the cerebral cortex. That's where information is processed, and thought produced. Then he creates ANNs to model each region. He's had success in motion and speech processing, shape detection, and other complex tasks. Now he's exploring how to computationally link his modules.

Machine-learning might have been a new concept to Good, but he would have encountered machine-learning algorithms in evaluating the Perceptron for IBM. Then, the tantalizing possibility of machines learning as humans do suggested to Good consequences others had not yet imagined. If a machine could make itself smarter, then the improved machine would be even better at making itself smarter, and so on.

In the tumultuous 1960s leading up to his creating the intelligence explosion concept, he already might have been thinking about the kinds of problems an intelligent machine could help with. There were no more hostile German U-boats to sink, but there was the hostile Soviet Union, the Cuban Missile Crisis, the assassination of President Kennedy, and the proxy war between the United States and China, fought across Southeast Asia. Man skated toward the brink of extinction—it seemed time for a new Colossus. In *Speculations*, Good wrote:

> [Computer pioneer] B. V. Bowden stated . . . that there is no point in building a machine with the intelligence of a man, since it is easier to construct human brains by the usual method . . . This shows that highly intelligent people can overlook the "intelligence explosion." It is true that it would be uneconomical to build a machine capable only of ordinary intellectual attainments, but it seems fairly probable that if this could be done then, at double the cost, the machine could exhibit ultraintelligence.

So, for a few dollars more you can get ASI, artificial superintelligence, Good proposes. But then watch out for the civilization-wide ramifications of sharing the planet with smarter than human intelligence.

In 1962, before he'd written "Speculations Concerning the First Ultraintelligent Machine," Good edited a book called *The Scientist Speculates*. He wrote a chapter entitled, "The Social Implications of Artificial Intelligence," kind of a warm-up for the superintelligence ideas he was developing. Like Steve Omohundro would argue almost fifty years later, he noted that among

the problems intelligent machines will have to address are those
caused by their own disruptive appearance on Earth.

> Such machines . . . could even make useful political
> and economic suggestions; and they would *need* to do
> so in order to compensate for the problems created by
> their own existence. There would be problems of over-
> population, owing to the elimination of disease, and of
> unemployment, owing to the efficiency of low-grade
> robots that the main machines had designed.

But, as I was soon to learn, Good had a surprising change of
heart later in life. I had always grouped him with optimists like
Ray Kurzweil, because he'd seen machines "save" the world before, and his essay hangs man's survival on the creation of a
superintelligent one. But Good's friend Leslie Pendleton had alluded to a turnabout. It took her a while to remember the occa-
sion, but on my last day in Blacksburg, she did.

In 1998, Good was given the Computer Pioneer Award of
the IEEE (Institute of Electrical and Electronics Engineers) Computer Society. He was eighty-two years old. As part of his accep-
tance speech he was asked to provide a biography. He submitted
it, but he did not read it aloud, nor did anyone else, during the
ceremony. Probably only Pendleton knew it existed. She included a copy along with some other papers I requested, and
gave them to me before I left Blacksburg.

Before taking on Interstate I-81, and heading back north, I
read it in my car in the parking lot of a Rackspace Inc. cloud
computing center. Like Amazon, and Google, Rackspace (corporate slogan: Fanatical Support®), provides massive computing
power for little money by renting time on its arrays of tens of

thousands of processors, and exabytes of storage space. Of course Virginia "Invent the Future" Tech would have a Rackspace facility at hand, and I wanted a tour, but it was closed. Only later did it seem eerie that a dozen yards from where I sat reading Good's biographical notes, tens of thousands of aircooled processors toiled away on the world's problems.

In the bio, playfully written in the third person, Good summarized his life's milestones, including a probably never before seen account of his work at Bletchley Park with Turing. But here's what he wrote in 1998 about the first superintelligence, and his late-in-the-game U-turn:

> [The paper] "Speculations Concerning the First Ultraintelligent Machine" (1965) . . . began: "The survival of man depends on the early construction of an ultraintelligent machine." Those were his [Good's] words during the Cold War, and he now suspects that "survival" should be replaced by "extinction." He thinks that, because of international competition, we cannot prevent the machines from taking over. He thinks we are lemmings. He said also that "probably Man will construct the *deus ex machina* in his own image."

I read that and stared dumbly at the Rackspace building. As his life wound down, Good had revised more than his belief in the probability of God's existence. I'd found a message in a bottle, a footnote that turned everything around. Good and I had something important in common now. We both believed the intelligence explosion wouldn't end well.

Chapter Eight
The Point of No Return

But if the technological Singularity can happen, it will. Even if all the governments of the world were to understand the "threat" and be in deadly fear of it, progress toward the goal would continue. In fact, the competitive advantage—economic, military, even artistic—of every advance in automation is so compelling that passing laws, or having customs, that forbid such things merely assures that someone else will.

—Vernor Vinge, The Coming Technological Singularity, 1993

This quotation sounds like a fleshed-out version of I. J. Good's biographical aside, doesn't it? Like Good, two-time Hugo Award-winning science fiction author and mathematics professor Vernor Vinge alludes to humans' lemminglike predilection to chase glory into the cannon's mouth, to borrow Shakespeare's phrase. Vinge told me he'd never read Good's self-penned biographical paragraphs, or learned about his late-in-life change of heart about the intelligence explosion. Probably only Good, and Leslie Pendleton, knew about it.

Vernor Vinge was the first person to formally use the word "singularity" when describing the technological future—he did it in a 1993 address to NASA, entitled "The Coming Technological Singularity." Mathematician Stanislaw Ulam reported that he and polymath John von Neumann had used "singularity" in a conversation about technological change thirty-five years earlier, in 1958. But Vinge's coinage was public, deliberate, and set the singularity ball rolling into the hands of Ray Kurzweil and what is today a Singularity movement.

With that street cred, why doesn't Vinge work the lecture and conference circuits as the ultimate Singularity pundit?

Well, singularity has several meanings, and Vinge's usage is more precise than others. To define singularity he made an analogy to the point in the orbit of a black hole beyond which light cannot escape. You can't see what's going on beyond that point, called the event horizon. Similarly, once we share the planet with entities more intelligent than ourselves, all bets are off—we cannot predict what will happen. You'd have to be at least that smart yourself to know.

So, if you don't have a good sense of how the future works out, how do you write about it? Vinge doesn't write science fantasy—he's what's known as a hard sci-fi author, using real science in his fiction. The singularity left him hamstrung.

It's a problem we face every time we consider the creation of intelligences greater than our own. When this happens, human history will have reached a kind of

singularity—a place where extrapolation breaks down
and new models must be applied—and the world will
pass beyond our understanding.

As Vinge tells it, when he began writing science fiction in
the 1960s, the science-based worlds he wrote about were forty
or fifty years away. But by the 1990s the future was running
toward him, and the rate of technological change seemed to be
accelerating. He could no longer anticipate what the future
would bring, because he reckoned it would soon contain greater-
than-human intelligence. That intelligence, and not mankind's,
would establish the rate of technological progress. He couldn't
write about it, and nor could others.

Through the sixties and seventies and eighties, recognition of the cataclysm spread. Perhaps it was the science-
fiction writers who felt the first concrete impact. After
all, the "hard" science-fiction writers are the ones who
try to write specific stories about all that technology may
do for us. More and more, these writers felt an opaque
wall across the future.

AI researcher Ben Goertzel told me, "Vernor Vinge saw its
inherent unknowability very clearly when he posited the notion of the technological singularity. It's because of that that he
doesn't go around giving speeches about it because he doesn't
know what to say. What's he going to say? 'Yeah I think we're
going to create technologies that will be much more capable
than humans and then who knows what will happen?'"

But what about the invention of fire, agriculture, the printing press, electricity? Haven't many technological "singularities"

already occurred? Disruptive technological change is nothing new, but no one felt compelled to come up with fancy names for its occurrences. My grandmother was born before automobiles were widely used, and lived to see Neil Armstrong walk on the moon. Her name for it was the twentieth century. What makes Vinge's transition so special?

"The secret sauce is intelligence," Vinge told me. His voice is a rapid-fire tenor, given to laughter. "Intelligence is what makes it different, and the defining operational feature is that the prior folk can't understand. We are in a situation that in a very brief time, just a few decades, we'll be getting transformations that are, by analogy, of biologically large significance."

Two important ideas are packed into this. First, the technological singularity will bring about a change in intelligence itself, the solely human superpower that creates technology to begin with. That's why it's different from any other revolution. Second, the biological transformation Vinge alludes to is when mankind took the world stage some two hundred thousand years ago. Because he was more intelligent than any other species, Homo sapiens, or "wise man," began to dominate the planet. Similarly, minds a thousand or a million times more intelligent than man's will change the game forever. What will happen to us?

This drew a percussive laugh from Vinge. "If I get pushed hard about questions about what the Singularity is going to be like, my most common retreat is to say, Why do you think I called it the singularity?"

But Vinge has concluded one thing about the opaque future—the Singularity is menacing, and could lead to our extinction. The author, whose 1993 speech quotes Good's 1967 intelligence explosion paragraph in its entirety, points out that the famous statistician didn't take his conclusions far enough:

Good has captured the essence of the runaway, but does not pursue its most disturbing consequences. Any intelligent machine of the sort he describes would not be humankind's "tool"—any more than humans are the tools of rabbits or robins or chimpanzees.

That's another apt analogy—rabbits are to humans as humans will be to superintelligent machines. And how do we treat rabbits? As pests, pets, or dinner. ASI agents will be our tools at first—their ancestors Google, Siri, and Watson are now. And, Vinge suggests, there are more ways besides stand-alone machine intelligence that could cause a singularity. They include intelligence emerging from the Internet, from the Internet *plus* its users (a digital Gaia), from human-computer interfaces, and from the biological sciences (improving the intelligence of future generations through gene manipulation).

In three of these routes, humans stay involved throughout the technologies' development, perhaps guiding a gradual and manageable intelligence enhancement rather than an explosion. So it's possible, Vinge says, to consider how mankind's greatest problems—hunger, disease, even death itself—may be conquered. That's the vision espoused by Ray Kurzweil and promulgated by "Singularitarians." Singularitarians are those who anticipate that mostly good things will emerge from the accelerated future. Their "singularity" sounds too rosy for Vinge.

"We're playing a very high-stakes game and the plus side of it is so optimistic that that by itself is sort of scary. A worldwide economic wind is associated with these advances in AI. And that is an extraordinary powerful force. So, there's hundreds of thousands of people in the world, very smart people, who are working on things that lead to superhuman intelligence. And probably most of them don't even look at it that way. They look at it as *faster, cheaper, better, more profitable.*"

Vinge compares it to the Cold War strategy called MAD— mutually assured destruction. Coined by acronym-loving John von Neumann (also the creator of an early computer with the winning initials, MANIAC), MAD maintained Cold War peace through the promise of mutual obliteration. Like MAD, superintelligence boasts a lot of researchers secretly working to develop technologies with catastrophic potential. But it's like mutually assured destruction without any commonsense brakes. No one will know who is ahead, so everyone will assume someone else is. And as we've seen, the winner won't take all. The winner in the AI arms race will win the dubious distinction of being the first to confront the Busy Child.

"We've got thousands of good people working all over the world in sort of a community effort to create a disaster," Vinge said. "The threat landscape going forward is very bad. We're not spending enough effort thinking about failure possibilities."

Some of the other scenarios Vinge is concerned about also warrant more attention. A digital Gaia, or marriage of humans and computers, is already organizing on the Internet. What that will mean for our future is profound and far-reaching, and deserving of more books than have been written about it. IA, or intelligence augmentation, has a similar potential for disaster as standalone AI, mitigated somewhat by the fact that a human takes part, at least at first. But that advantage will quickly disappear. We'll talk more about IA later. First, I want to pay attention to Vinge's notion that intelligence could emerge from the Internet.

Technology thinkers, including George Dyson and Kevin

Kelly, have proposed that information is a life-form. The computer code that carries information replicates itself and grows according to biological rules. But intelligence, well that's something else. It's a feature of complex organisms, and it doesn't come by accident.

At his home in California, I had asked Eliezer Yudkowsky if intelligence could emerge from the exponentially growing hardware of the Internet, from its five trillion megabytes of data, its more than seven billion connected computers and smart phones, and its seventy-five million servers. Yudkowsky had grimaced, as if his brain cells had been flooded by an acid bath of dumb.

"Flatly, no," he said. "It took billions of years for evolution to cough up intelligence. Intelligence is not emergent in the complexity of life. It doesn't happen automatically. There is optimization pressure with natural selection."

In other words, intelligence doesn't arise from complexity alone. And the Internet lacks the kinds of environmental pressures that in nature favored some mutations over others.

"I have a saying that there's probably less interesting complexity in the entire Milky Way outside Earth than there is in one earthly butterfly because the butterfly has been produced by processes that retain their successes and build on them," Yudkowsky said.

I agree that intelligence wouldn't spontaneously blossom from the Internet. But I think agent-based financial modeling could soon change everything about the Net itself.

Once upon a time when Wall Street analysts wanted to predict how the market would behave, they turned to a series of rules prescribed by macroeconomics. These rules consider factors like interest rates, employment data, and "housing starts" or new houses built. Increasingly, however, Wall Street has turned to agent-based financial modeling. This new science can computationally simulate the entire stock market, and even the economy, to improve forecasting.

To model the market, researchers make computer models of the entities buying and selling stocks—individuals, firms, banks, hedge funds, and so on. Each of these thousands of "agents" has different goals and decision rules, or strategies, for buying and selling. They in turn are influenced by ever-changing market data. The agents, powered by artificial neural networks and other AI techniques, are "trained" on real-world information. Acting in unison, and updated with live data, the agents create a fluid portrait of the living market.

Then, analysts test scenarios for trading individual securities. And through evolutionary programming techniques, the market model can "step forward" a day or a week, giving analysts a good idea of what the market will look like in the future, and what investment opportunities might appear. This "bottom-up" approach to creating financial models embodies the idea that simple behavioral rules of individual agents generate complex overall behavior. Generally speaking, what's true of Wall Street is also true of beehives and ant colonies.

What begins to take shape in the supercomputers of the financial capitals of the world are virtual worlds steeped in real-world detail, and populated by increasingly intelligent "agents." Richer, more nuanced forecasting equals bigger profits. So powerful economic incentives fuel the drive for increasing the models' precision at every level.

If it's useful to create computational agents that exercise complex stock-buying strategies, wouldn't it be *more* useful to create computational models with the full range of human motivations and abilities? Why not create AGIs, or human-level intelligent agents? Well, that's what Wall Street is doing, but by another name—agent-based financial models.

That financial markets will give rise to AGI is the position of Dr. Alexander D. Wissner-Gross. Wissner-Gross has the kind of résumé that makes other inventors, scholars, and polymaths linger near open elevator shafts. He's authored thirteen publications, holds sixteen patents, matriculated with a triple major from MIT in physics, electrical science and engineering, and mathematics, while graduating first in his class at MIT's School of Engineering. He's got a Ph.D. in physics from Harvard and won a big prize for his thesis. He has founded and sold companies, and according to his résumé, won "107 major distinctions," which are probably not of the "employee of the week" variety. He's now a Harvard research fellow, trying to commercialize his ideas about computational finance.

And he thinks that while brilliant theorists around the world are racing to create AGI, it might appear fully formed in the financial markets, as an unintended consequence of creating computational models of large numbers of humans. Who'd create it? "Quants," Wall Street's name for computer finance geeks.

"It's certainly possible that a real living AGI could emerge on the financial markets," Wissner-Gross told me. "Not the result of a single algorithm from a single quant, but an aggregate of all the algorithms from lots of hedge funds. AGI may not need a coherent theory. It could be an aggregate phenomenon. If you follow the money, finance has a decent shot at being the primordial ooze out of which AGI emerges."

To buy this scenario you have to believe that there's a lot of money firing the creation of better and better financial modeling. And in fact there is—anecdotally, more money than anyone else is spending on machine intelligence, perhaps even more than DARPA, IBM, and Google can throw at AGI. That translates into more and better supercomputers and smarter quants. Wissner-Gross said quants use the same tools as AI researchers—neural nets, genetic algorithms, automatic reading, hidden Markov models, you name it. Every new AI tool gets tested in the crucible of finance.

"Whenever a new AI technique is developed," Wissner-Gross told me, "the first question out of anyone's mouth is 'Can you use it to trade stocks?'"

Now imagine you're a high-powered quant with a war chest big enough to hire more quants and buy more hardware. The hedge fund you work for is running a great model of the Street, populated with thousands of diverse economic agents. Its algorithms react with those of other hedge funds—they're so tightly coupled that they rise and fall together, seeming to act in concert. According to Wissner-Gross, market observers have suggested that some seem to be *signaling* each other across Wall Street with millisecond trades that occur at a pace no human can track (these are HFTs or high-frequency trades, discussed in chapter 6).

Wouldn't the next logical step be to make your hedge fund reflective? That is, perhaps your algorithm shouldn't automatically trigger sell orders based on another fund's massive sell-off (which is what happened in the flash crash of May 2010). Instead it would perceive the sell-off and see how it was impacting other funds, and the market as a whole, before making its move. It might make a different, better move. Or maybe it could do one better, and simultaneously run a very large number of hypothetical markets, and be prepared to execute one of many strategies in response to the right conditions.

In other words, there are huge financial incentives for your algorithm to be self-aware—to know exactly what it is and model the world around it. That is a *lot* like AGI. This is without a doubt the way the market is headed, but is anyone cutting to the chase and building AGI?

Wissner-Gross didn't know. And he might not tell you if he did. "There are strong mercantile impulses to keep secret any seriously profitable advances," he said.

Of course. And he's not just talking about competition among hedge funds, but a kind of natural selection among algorithms. Winners thrive and pass on their code. Losers die. The market's evolutionary pressure would speed the development of intelligence, but not without a human quant's guiding hand. Yet.

And an intelligence explosion would be opaque in the computational finance universe, for at least four reasons. Like many cognitive architectures, it would probably use neural networks, genetic programming, and other "black box" AI techniques. Second, the high-bandwidth, millisecond-fast transmissions take place faster than humans can react—look at what happened during the Flash Crash. Third, the system is incredibly complex—there's no one quant or even a group of quants (a quantum? a gaggle? what's the quants' collective noun?) who can explain the algorithm ecosystem of Wall Street and how the algorithms interact.

Finally, if a formidable intelligence emerged from computational finance, it would almost certainly be kept secret so long as it was making money for its creators. That's four levels of opacity.

To sum up, AGI could arise from Wall Street. The most successful algorithms are kept secret by the quants who lovingly code them, or the companies that own them. An intelligence explosion would be invisible to most if not all humans, and probably unstoppable anyway.

The similarities between computational finance and AGI research don't end there. Wissner-Gross has another astonishing proposal. He claims that the first strategies to control AGI might arise from measures now proposed to control high-frequency trading. Some of them sound promising.

Market circuit breakers would cut off hedge fund AIs from the outside world, in case of emergency. They'd detect cascading algorithm interactions like the 2010 Flash Crash and unplug the machines.

The Large Trader Rule requires detailed registration of AIs, along with full human organization charts. If this sounds like a prelude to large government intervention, it is. Why not? Wall Street has proven again and again that as a culture it cannot behave responsibly without strenuous regulation. Is that also true of AGI developers? Without a doubt. There's no moral merit badge required for studying AGI.

Pre-trade testing of algorithms could simulate algorithms' behavior in a virtual environment before they were let loose on the market. *AI Source Code audits* and *Centralized AI Activity Recording* aim to anticipate errors, and facilitate after-game analysis following an accident, like the 2010 Flash Crash.

But look back at the four levels of opacity mentioned earlier, and see if these defenses, even if they were fully implemented, sound anything like foolproof to you.

As we've seen, Vinge took the baton from I. J. Good and gave the intelligence explosion important new attributes. He considered alternate routes to achieving it besides the neural nets Good anticipated, and pointed out the possibility, even probability, of human annihilation. Most important, perhaps, Vinge gave it a name—a singularity.

Naming things, as Vinge, author of the seminal science-fiction novella *True Names* well knows, is a powerful act. Names stick on the lips, lodge in the brain, and hitchhike across generations. In the book of Genesis, theologians propose, naming everything on Earth on the seventh day was important because a rational creature was about to share the stage God made, and would make use of names thereafter. Lexical growth is an important part of childhood development—without language the brain doesn't develop normally. It seems unlikely that AGI will be possible without language, without nouns, without names.

Vinge named the singularity to designate a scary place for humans to be, an unsafe proposition. His definition of the singularity is metaphorical—the orbit outside a black hole where gravitational forces are so strong not even light can escape. We cannot know its essence, and it was named that way on purpose.

Then suddenly, all that changed.

To the idea of a singularity as espoused by Vinge, Ray Kurzweil added a dramatic catalyst that shifts this whole conversation into hyperdrive, and brings into sharper focus the catastrophic danger ahead: the exponential growth of computer power and speed. It's because of this growth that you should cast a jaundiced eye on anyone who claims human-level machine intelligence won't be achieved for a century or more, if at all.

Per dollar spent, computers have increased in power by a billion times in the last thirty years. In about twenty years a thousand dollars will buy a computer a million times more powerful than one today, and in twenty-five years a *billion* times more powerful than one today. By about 2020 computers will be able to model the human brain, and by 2029 researchers will be able to run a brain simulation that's every bit as intellectually and emotionally nuanced as a human mind. By 2045, human and machine intelligence will have increased a *billionfold*, and will develop technologies to defeat our human frailties, such as fatigue, illness, and death. In the event we survive it, the twenty-first century won't see a century's worth of technological progress, but 200,000 years' worth.

This juggernaut of projections and analysis is Kurzweil's, and it's the key to understanding the third and reigning definition of the singularity—his. It's at the heart of Kurzweil's Law of Accelerating Returns, a theory about technological progress that Kurzweil didn't invent, but pointed out, in much the same way Good anticipated the intelligence explosion and Vinge warned of a coming singularity. What the Law of Accelerating Returns means is that the projections and advances we're discussing in this book are hurtling toward us like a freight train that doubles its speed every mile, then doubles again. It's very hard to perceive how quickly it will get here, but suffice it to say that if that train were traveling twenty miles an hour by the end of the first mile, just fifteen miles later it'll be traveling more than 65,000 miles an hour. And, it's important to note Kurzweil's projections aren't just about advances in technology hardware, such as what's inside a new iPhone, but advances in the technology arts, like developing a unified theory of artificial intelligence.

But here's where Kurzweil and I differ. Instead of leading to a kind of paradise, as Kurzweil's aggregate projections assert, I believe the Law of Accelerating Returns describes the shortest possible distance between our lives as they are and the end of the human era.

Chapter Nine

The Law of Accelerating Returns

Computing is undergoing the most remarkable transformation since the invention of the PC. The innovation of the next decade is going to outstrip the innovation of the past three combined.

— Paul Otellini, CEO of Intel

With his books *The Age of Spiritual Machines: When Computers Exceed Human Intelligence* and *The Singularity Is Near*, Ray Kurzweil commandeered the word "singularity" and changed its meaning to that of a bright, hopeful period of human history, which his tools of extrapolation let him see with remarkable precision. Sometime in the next forty years, he writes, technological development will advance so rapidly that human existence will be fundamentally altered, the fabric of history torn. Machines and biology will become indistinguishable. Virtual worlds will be more vivid and captivating than reality. Nanotechnology will enable manufacturing on demand, ending hunger and poverty, and delivering cures for all of mankind's diseases. You'll be able to stop your body's aging, or even reverse it. It's the most important time to be alive not just because you will witness a truly stupefying pace of technological transformation, but because the technology promises to give you the tools to live forever. It's the dawn of a "singular" era.

> What, then, is the Singularity? It's a future period during which the pace of technological change will be so rapid, its impact so deep, that human life will be irreversibly transformed. Although neither utopian or dystopian, this epoch will transform the concepts that we rely on to give meaning to our lives, from our business models to the cycle of human life, including death itself . . .
>
> Consider J. K. Rowling's Harry Potter stories from this perspective. These tales may be imaginary, but they are not unreasonable visions of our world as it will exist only a few decades from now. Essentially all of the Potter "magic" will be realized through the technologies I will explore in this book. Playing quidditch and transforming people and objects into other forms will be feasible in full-immersion virtual-reality environments, as well as in real reality, using nanoscale devices.

So, the singularity will be "neither utopian or dystopian" but we'll get to play quidditch! Obviously, Kurzweil's Singularity is dramatically different from Vernor Vinge's singularity and I. J. Good's intelligence explosion. Can they be reconciled? Is it simultaneously the best time to be alive, and the worst? I've read almost every word Kurzweil has published, and listened to every available audio recording, podcast, and video. In 1999 I interviewed him at length for a documentary film that was in part about AI. I know what he's written and said about the dangers of AI, and it isn't much.

Surprisingly, however, he was indirectly responsible for the subject's most cogent cautionary essay—Bill Joy's "Why the Future Doesn't Need Us." In it, Joy, a programmer, computer architect, and the cofounder of Sun Microsystems, urges a slowdown and even a halt to the development of three technologies he believes are too deadly to pursue at the current pace: artificial intelligence, nanotechnology, and biotechnology. Joy was prompted to write it after a frightening conversation in a bar with Kurzweil, followed by his reading *The Age of Spiritual Machines.* In nonscholarly literature and lectures about the perils of AI, I think only Asimov's Three Laws are cited more frequently, albeit misguidedly, than Joy's hugely influential essay. This paragraph sums up his position on AI:

> But now, with the prospect of human-level computing power in about thirty years, a new idea suggests itself: that I may be working to create tools which will enable the construction of the technology that may replace our species. How do I feel about this? Very uncomfortable. Having struggled my entire career to build reliable software systems, it seems to me more than likely that this future will not work out as well as some people may have imagined. My personal experience suggests we tend to overestimate our design abilities. Given the incredible power of these new technologies, shouldn't we be asking how we can best coexist with them? And if our own extinction is a likely, or even possible, outcome of our technological development, shouldn't we proceed with great caution?

Kurzweil's bar talk can start a national dialogue, but his few cautionary words get lost in the excitement of his predictions. He insists he's not painting a utopian view of tomorrow, but I don't think there's any doubt that he is.

Few write more knowledgably or persuasively about technology than Kurzweil—he takes pains to make himself clearly understood, and he defends his message with humility. However, I think he has made a mistake by appropriating the name "singularity" and giving it a new, rosy meaning. So rosy that I, like Vinge, find the definition scary, full of compelling images and ideas that mask its danger. His rebranding underplays AI's peril and overinflates the promise. Starting from a technological proposition, Kurzweil has created a cultural movement with strong religious overtones. I think mixing technological change and religion is a big mistake.

> Imagine a world where the difference between man and machine blurs, where the line between humanity and technology fades, and where the soul and the silicon chip unite . . . In [Kurzweil's] inspired hands, life in the new millennium no longer seems daunting. Instead, Kurzweil's twenty-first century promises to be an age in which the marriage of human sensitivity and artificial intelligence fundamentally alters and improves the way we live.

Kurzweil's not just the godfather of Singularity issues, a polite, relentless debater, and a tireless if rather mechanical promoter. He's the den-master for a lot of young men, and some women, living on the singularity edge. Singularitarians tend to be twenty- and thirty-somethings, male, and childless. For the most part, they're smart white guys who've heard the call of the Singularity. Many have answered by dropping the kinds of careers that would've made their parents proud to take on monkish lives committed to Singularity issues. A lot are autodidacts, probably in part because no undergraduate program offers a major in computer science, ethics, bioengineering, neuroscience, psychology, and philosophy, in short, Singularity studies. (Kurzweil cofounded Singularity University, which offers no degrees and isn't accredited. But it promises "a broad, cross-disciplinary understanding of the biggest ideas and issues in transformative technologies.") Many Singularitarians are too smart and self-directed to get in line for traditional education anyway. And many are addled wing nuts few colleges or universities would invite on campus.

Some Singularitarians have adopted rationality as the major tenet of their creed. They believe that greater logical and reasoning abilities, particularly among tomorrow's decision makers, decreases the probability that we'll commit suicide by AI. Our brains, they argue, are full of bizarre biases and heuristics that served us well during our evolution, but get us into trouble when confronted with the modern world's complex risks and choices. Their main focus isn't on a catastrophic, negative Singularity, but a blissful, positive one. In it, we can take advantage of life-extending technologies that let us live on and on, probably in mechanical rather than biological form. In other words, cleanse yourself of faulty thinking, and you can find deliverance from the world of the flesh, and discover life everlasting.

It's no surprise that the Singularity is often called the Rapture of the Geeks—as a movement it has the hallmarks of an apocalyptic religion, including rituals of purification, eschewing frail human bodies, anticipating eternal life, and an uncontested (somewhat) charismatic leader. I wholeheartedly agree with the Singularitarian idea that AI is the most important thing we could be thinking about right now. But when it comes to immortality talk, I get off the bus. Dreams about eternal life throw out a powerful distortion field. Too many Singularitarians believe that the confluence of technologies presently accelerating will not yield the kinds of disasters we might anticipate from any of them individually, nor the conjunctive disasters we might also foresee, but instead will do something 180 degrees different. It will save mankind from the thing it fears most. Death.

But how can you competently evaluate tools, and whether and how their development should be regulated, when you believe the same tools will permit you to live forever? Not even the world's most rational people have a magical ability to dispassionately evaluate their own religions. And as scholar William Grassie argues, when you are asking questions about transfiguration, a chosen few, and living forever, what are you taking about if not religion?

> Will the Singularity lead to the supersession of humanity by spiritual machines? Or will the Singularity lead to the transfiguration of humanity into superhumans who live forever in a hedonistic, rationalist paradise? Will the Singularity be preceded by a period of tribulation? Will there be an elect few who know the secrets of the Singularity, a vanguard, perhaps a remnant who make it to the Promised Land? These religious themes are all present in the rhetoric and rationalities of the Singularitarians, even if the pre- and post-millennialist in-

terpretations aren't consistently developed, as is certainly the case with pre-scientific Messianic movements.

Unlike Good's and Vinge's takes on the accelerating future, Kurzweil's Singularity isn't brought about by artificial intelligence alone, but by three technologies advancing to points of convergence—genetic engineering, nanotechnology, and robotics, a broad term he uses to describe AI. Also unlike Good and Vinge, Kurzweil has come up with a unified theory of technological evolution that, like any respectable scientific theory, tries to account for observable phenomena, and makes predictions about future phenomena. It's called the Law of Accelerating Returns, or LOAR.

First, Kurzweil proposes that a smooth exponential curve governs evolutionary processes, and that the development of technology is one such evolutionary process. Like biological evolution, technology evolves a capability, then uses that capability to evolve to the next stage. In humans, for instance, big brains and opposable thumbs allowed toolmaking and the power grip needed to use our tools effectively. In technology, the printing press contributed to bookbinding, literacy, and to the rise of universities and more inventions. The steam engine leveraged the Industrial Revolution and more and more inventions.

Because of its way of building upon itself, technology starts slow, but then its growth curve steepens until it shoots upward almost vertically. According to Kurzweil's trademark graphs and charts, we are entering the most critical period of technological evolution, the steep upward part, the "knee of the exponential curve." It's all up from here.

Kurzweil developed his Law of Accelerating Returns to describe the evolution of any process in which patterns of information evolve. He applies LOAR to biology, which favors increasing molecular order, but it is more convincing when used to anticipate the pace of change in information technologies, including computers, digital cameras, the Internet, cloud computing, medical diagnostic and treatment equipment, and more—any technology involved in the storage and retrieval of information.

As Kurzweil notes, LOAR is fundamentally an economic theory. Accelerating returns are fueled by innovation, competition, market size—the features of the marketplace and manufacturers. In the computer market the effect is described by Moore's law, another economic theory disguised as a technology theory, first observed in 1965 by Intel's cofounder Gordon Moore.

Moore's law states that the number of transistors that can be put on an integrated circuit to build a microprocessor doubles every eighteen months. A transistor is an on/off switch that can also amplify an electrical charge. More transistors equals more processing speed, and faster computers. Moore's law means computers will get smaller, more powerful, and cheaper at a reliable rate. This does not happen because Moore's law is a natural law of the physical world, like gravity, or the Second Law of Thermodynamics. It happens because the consumer and business markets motivate computer chip makers to compete and contribute to smaller, faster, cheaper computers, smart phones, cameras, printers, solar arrays, and soon, 3-D printers. And chip makers are building on the technologies and techniques of the past. In 1971, 2,300 transistors could be printed on a chip. Forty years, or twenty doublings later, 2,600,000,000. And with those transistors, more than two million of which could fit on the period at the end of this sentence, came increased speed.

Here's a dramatic case in point. Jack Dongarra, a researcher at Tennessee's Oak Ridge National Lab and part of a team that tracks supercomputer speed, determined that Apple's best-selling tablet, the iPad 2, is as fast as a circa 1985 Cray 2 supercomputer. In fact, running at over 1.5 gigaflops (one gigaflop equals one *billion* mathematical operations, or calculations per second) the iPad 2 would have made the list of the world's five hundred fastest supercomputers as late as 1994.

In 1994, who could have imagined that less than a generation later a supercomputer smaller than a textbook would be economical enough to be given free to high school students, and what's more, it would connect to the sum of mankind's knowledge, sans cables? Only Kurzweil would've been so bold, and while he didn't make this exact claim about supercomputers, he did anticipate the Internet's explosion.

In information technologies, each breakthrough pushes the next breakthrough to occur more quickly—the curve we talked about gets steeper. So when considering the iPad 2 the question isn't what we can expect in the *next* fifteen years. Instead, look out for what happens in a fraction of that time. By about 2020, Kurzweil estimates we'll own laptops with the raw processing power of human brains, though not the intelligence.

Let's have a look at how Moore's law may apply to the intelligence explosion. If we assume AGI can be attained, Moore's law implies the recursive self-improvement of an intelligence explosion may not even be necessary to achieve ASI, or superhuman intelligence. That's because once you've achieved AGI, less than two years later machines of human-level intelligence will have doubled in speed. Then in under two years, another doubling. Meanwhile, average human intelligence will remain the same. Soon the AGI has left us behind.

If the intelligence starts taking part in its own modification, then what happens? Eliezer Yudkowsky describes how quickly after AGI the pace of technological progress may slip from our hands.

If computing speeds double every two years, what happens when computer-based AIs are doing the research?

> Computing speed doubles every two years.
> Computing speed doubles every two years of work.
> Computing speed doubles every two subjective years of work.
> Two years after Artificial Intelligences reach human equivalence, their speed doubles.
> One year later, their speed doubles again. Six months—three months—1.5 months . . .
> Singularity.

Some object that Moore's law will stop before 2020, when it becomes physically impossible to fit more transistors onto an integrated circuit. Others think Moore's law will give way to *faster* doublings when processors undergo technological make-overs, using smaller components to perform computations, such as atoms, photons of light, even DNA. 3-D processor chips developed by Switzerland's École Polytechnique Fédérale de Lausanne could be the first to beat Moore's law. Though not yet in production, EPFL's processor chips are stacked vertically instead of horizontally, and will be faster and more efficient than traditional chips, as well as parallel-processing ready. And already the company Gordon Moore cofounded may have bought his law more time with the creation of the first 3-D *transistors*. Recall that transistors are electrical switches. Traditional transistors work by regulating electrical current moving across two dimensions. Intel's new Tri-Gate transistors conduct current over three dimensions, for a 30 percent increase in speed, and up to 50 percent savings in power. A billion Tri-Gate transistors will be on each of Intel's next line of processor chips.

Because imprinting transistors on silicon is involved in many information technologies, from cameras to medical sensors, Moore's law applies to them, too. But Moore was theorizing about integrated circuits, not the many linked worlds of information technology that include both products and processes. So, Kurzweil's more general law, the Law of Accelerating Returns, is a better fit. And more technologies are *becoming* information technologies, as computers, and even robots, grow ever more intimately involved with every aspect of product design, manufacture, and sales. Consider that every smart phone's manufacture—not just its processor chip—took advantage of the digital revolution. It's been just six years since Apple's iPhone first came out, and Apple has released *six* versions. Apple has more than doubled its speed and for most users halved its price, or better. That's because hardware speed has been regularly doubling in the components within the end product. But it's also been doubling in every link in the production pipeline that led to its creation.

The effects anticipated by LOAR reach far beyond the computer and smart phone businesses. Recently Google's cofounder Larry Page met with Kurzweil to discuss global warming, and they parted optimistic. In twenty years, they claimed, nanotechnology will enable sun-powered energy to become more economical than oil or coal. The industry will be able to provide the earth with 100 percent of its energy needs. While solar energy supplies just a half a percent of the world's energy requirements today, they reason that rate is doubling every two years as it has for the last twenty years. So, in two more years solar energy will account for 1 percent of world energy requirements, in four years, 2 percent, and in sixteen more years, eight more doublings, or two to the eighth power, equaling 256 percent of world energy needs. Even accounting for increased population and energy demands two decades hence, that should be enough solar power to cover it and then some. And so, according to Kurzweil and Page, global warming will be solved.

And so will, um, mortality. According to Kurzweil, the means are almost in reach for extending life indefinitely.

"We now have the actual means of understanding the software of life and reprogramming it; we can turn genes off without any interference, we can add new genes, whole new organs with stem cell therapy," Kurzweil said. "The point is that medicine is now an information technology—it's going to double in power every year. These technologies will be a million times more powerful for the same cost in twenty years."

Kurzweil believes that the shortest route to AGI is to reverse engineer the brain—intricately scanning it to yield a collection of brain-based circuits. Represented in algorithms or hardware networks, these circuits will then be fired up on a computer as a unified synthetic brain, and taught everything it needs to know. Several organizations are working on projects to accomplish this path to AGI. We'll discuss some approaches and roadblocks ahead.

The evolutionary pace of the hardware needed to run a virtual brain calls for a closer look. Let's start with the human brain and work our way toward computers that could emulate it. Kurzweil writes that the brain has about 100 billion neurons, each connected to about a thousand other neurons. That makes about 100 trillion interneuronal connections. Each connection is capable of making some 200 calculations per second (electronic circuits are at least 10 million times faster). Kurzweil multiplies the brain's interneuronal connections by their calculations per second and gets 20 million billion calculations per second, or 20,000,000,000,000,000.

The title of fastest supercomputer changes hands almost monthly, but right now the Department of Energy's Sequoia reigns with more than sixteen petaflops. That's 16,000,000,000,000,000 calculations per second, or roughly 80 percent of the speed of the human brain, as calculated by Kurzweil in 2000. But by 2005's *The Singularity Is Near*, Kurzweil had trimmed his brain-speed calculation down from twenty to sixteen petaflops, and estimated a supercomputer would reach it by 2013. Sequoia achieved it a year sooner.

Are we that close to brute-forcing brainpower? The numbers can be deceiving. Brains are parallel processors and excel at some jobs, while computers operate serially and excel at others. Brains are slow and work in spikes of neuronal activity. Computers can process faster and for longer, even indefinitely.

But human brains remain our sole example of advanced intelligence. If "brute force" can compete, computers will have to perform impressive cognitive feats. But consider a few of the complex systems today's supercomputers routinely model: weather systems, 3-D nuclear detonations, and molecular dynamics for manufacturing. Does the human brain contain a similar magnitude of complexity, or an order of magnitude higher? According to all indications, it's in the same ballpark.

Perhaps, as Kurzweil says, conquering the human brain is just around the corner, and the next thirty years of computer science will be like a hundred and forty at today's rate of progress. Factor in too that *creating* AGI is also an information technology. With exponentially increased computer speed, AI researchers can conduct their work faster. That means writing more complex algorithms, more processing-intensive algorithms, taking on harder computational problems, and conducting more experiments. Faster computers contribute to a more robust AI industry, which in turn produces more computer researchers, and faster, more useful tools for achieving AGI.

Kurzweil writes that after researchers produce a computer capable of passing the Turing test by 2029, things will accelerate even faster. But he doesn't predict a full-blown Singularity until sixteen years later, or 2045. Then the pace of technological advance will exceed our brains' ability to steer it. Therefore, he argues, we must augment them in order to keep up. That means plugging brain-boosting technologies directly into our neuro-circuitry, in the same way that today's cochlear implants connect to auditory nerves to help the hearing impaired. We'll perk up those slow interneuronal connections, and think faster, more deeply, while remembering more. We'll have access to all of human knowledge, and will, computerlike, instantly be able to share our thoughts and experiences with others, while experiencing theirs. Ultimately technology will enable us to upgrade our brains to a medium more durable than brain tissue, or we will upload our minds to computers, all the while preserving the qualities that make us *us*.

This outline of the future assumes of course that the *you* in you, your self, is transportable, and that's one whale of an assumption. But for Kurzweil, this is the path to immortality, and a breadth of knowledge and experience beyond what we can currently comprehend. Intelligence augmentation will happen so incrementally that few will reject it. But "incrementally" means by 2045, so over roughly the next thirty years, with the vast majority of change taking place in the last half decade or so. Is that gradual? I don't think so.

As we noted earlier, Apple came out with six versions of the iPhone in six years. According to Moore's law, their hardware was sufficiently advanced to undergo two or more doublings in that time, yet it underwent just one. Why? Because of the lag time taken by development, prototyping, and manufacture of the iPhone's components, including its processor, camera, memory, storage, screen, and so on, followed by the marketing and sale of the iPhone itself.

Will the lag time from marketing to sales ever go away? Perhaps someday hardware, like software, will upgrade itself automatically. But that probably won't happen until science has mastered nanotechnology or 3-D printing becomes ubiquitous. And when we're upgrading components of our own brains, instead of updating Microsoft Office or buying a few chips of RAM, it'll be a much more delicate procedure than anything we've experienced before, at least at first.

Yet Kurzweil claims that in this century we'll experience 200,000 years of technological progress in a hundred calendar years. Could we tolerate so much progress coming so fast?

Nicholas Carr, author of *The Shallows*, argues that smart phones and computers are lowering the quality of our thoughts, and changing the shape of our brains. In his book, *Virtually You*, psychiatrist Elias Aboujaoude warns that social networking and role-playing games encourage a swarm of maladies, including narcissism and egocentricity. Immersion in technology weakens individuality and character, proposes the programmer who *pioneered* virtual reality, Jaron Lanier, author of *You Are Not a Gadget: A Manifesto*. These experts warn that detrimental effects come from computers *outside* our bodies. Yet Kurzweil proposes only good things will come of computers *inside* our bodies. I think it's implausible to expect that hundreds of thousands of years of evolution will turn on a dime in thirty years, and that we can be reprogrammed to love an existence that is so different from the lives we've evolved to fit.

It's more likely that humans will decide on a rate of change they can manage and control. Each may choose differently, with many people settling on similar rates just as they settle on similar clothing styles, cars, and computers. We know Moore's law and LOAR are economic rather than deterministic laws. If enough people with enough resources want to artificially accelerate their brains, they'll create *some* demand. However, I think Kurzweil is greatly overestimating future persons' commitment to faster thinking and longer living. I don't think a Singularity full of delights, as he defines it, will be available anyway. AI developed without sufficient safeguards will prevent it.

The quest to create AGI is unstoppable and probably ungovernable. And because of the dynamics of doublings expressed by LOAR, AGI will take the world stage (and I mean *take*) much sooner than we think.

Chapter Ten

The Singularitarian

In contrast with our intellect, computers double their performance every eighteen months. So the danger is real that they could develop intelligence and take over the world.

— Stephen Hawking, physicist

Within thirty years, we will have the technological means to create superhuman intelligence. Shortly after, the human era will be ended. Is such progress avoidable? If not to be avoided, can events be guided so that we may survive?

— Vernor Vinge, author, professor, computer scientist

Each year since 2005, the Machine Intelligence Research Institute, formerly the Singularity Institute for Artificial Intelligence, has held a Singularity Summit. Over two days, a roster of speakers preach to about a thousand members of the choir about the Singularity big picture—its impact on jobs and the economy, health and longevity, and its ethical implications. Speakers at the

2011 summit in New York City included science legends, like Mathematica's Stephen Wolfram, Peter Thiel, a dot-com billionaire who pays tech-savvy teens to skip college and start companies, and IBM's David Ferrucci, principal investigator for the DeepQA/Watson Project. Eliezer Yudkowsky always speaks, and there's usually an ethicist or two as well as spokespeople for the extropian and transhuman communities. Extropians explore technologies and therapies that will permit humans to live forever. Transhumans think about hardware and cosmetic ways for increasing human capability, beauty, and . . . opportunities to live forever. Standing astride all the factions is the Colossus of the Singularity, a cofounder of the Singularity Summits, and the star of each gathering, Ray Kurzweil.

The 2011 summit's theme was IBM's DeepQA (question answering) computer Watson, and Kurzweil gave a rote presentation on the history of chatbots and Q&A systems, entitled "From Eliza to Watson." But midtalk he came to life to gut a maladroit essay coauthored by Microsoft cofounder Paul Allen, attacking his Singularity hypothesis.

Kurzweil didn't look particularly well—thin, a little halting, quieter than usual. He's not the kind of speaker who eats up the stage or delivers zingers. To the contrary, he's got a mild, robotic delivery, the kind that seems designed for hostage negotiations or bedtime stories. But it plays well against the casually revolutionary things he has to say. In an age when dot-com billionaires give presentations in pressed jeans, Kurzweil is wearing your uncle's brown slacks, along with tasseled loafers, sports jacket, and glasses. He's neither large nor small, but recently has started looking old, especially in comparison to the vigorous Kurzweil of my memory. He had been a mere fifty-two or so when I last interviewed him, and hadn't yet started the intense calorie restriction diet that's now part of his plan to slow his aging. With a fine-tuned regimen of diet, exercise, and supplements, Kurzweil plans to fend off death until technology finds the cure he's certain will come.

"I'm an optimist. As an inventor I have to be."

After his talk, Kurzweil and I sat knee to knee on metal chairs in a small dressing room a floor above the stage. A documentary film crew waited outside to get time with him after me. Just a decade before, when he had been a semifamous inventor and author, I'd monopolized him for three enjoyable hours with a film crew of my own; now he was a one-man industry whose attention I'd have as long as I could keep the door shut. I, too, was different—at our first meeting I had been gobsmacked by the idea of porting my brain to a computer, as described in *Spiritual Machines*. My questions were as tough as champagne bubbles. Now I'm more cynical, and wise to the dangers that no longer interest the master himself.

"I discuss quite a bit about peril in *The Singularity Is Near*," Kurzweil protested when I asked if he'd overspun the Singularity's promise and undersold its dangers. "Chapter eight is the deeply intertwined promise *and* peril in GNR [genetics, nanotechnology, and robotics] and I go into pretty graphic detail on the downsides of those three areas of technology. And the downside of robotics, which really refers to AI, is the most profound because intelligence is the most important phenomenon in the world. Inherently there is no absolute protection against strong AI."

Kurzweil's book does underline the dangers of genetic engineering and nanotechnology, but it gives only a couple of anemic pages to strong AI, the old name for AGI. And in that chapter he also argues that relinquishment, or turning our backs on some technologies because they're too dangerous, as advocated by Bill Joy and others, isn't just a bad idea, but an immoral one. I agree relinquishment is unworkable. But immoral?

"Relinquishment is immoral because it would deprive us of profound benefits. We'd still have a lot of suffering that we can overcome and therefore have a moral imperative to do that. Secondly, relinquishment would require a totalitarian system to ban the technology. And thirdly and most importantly it wouldn't work. It would just drive the technologies underground where irresponsible practitioners would then have no limitations. Responsible scientists who are charged with developing these defenses would not have access to the tools to do that. And it would actually be more dangerous."

Kurzweil is criticizing what's called the Precautionary Principle, a proposition that came out of the environmental movement, which, like relinquishment, is a straw man in this conversation. But it's important to spell out the principle, and see why it doesn't carry weight. The Precautionary Principle states, "If the consequences of an action are unknown but judged by some scientists to have even a small risk of being profoundly negative, it's better to not carry out the action than risk negative consequences." The principle isn't frequently or strictly applied. It would halt any purportedly dangerous technology if "some scientists" feared it, even if they couldn't put their finger on the causal chain leading to their feared outcome.

Applied to AGI, the Precautionary Principle and relinquishment are nonstarters. Barring a catastrophic accident on the way to AGI that would scare us straight, both measures are unenforceable. The best corporate and government AGI projects will seek the competitive advantage of secrecy—we have seen it already in stealth companies. Few countries or corporations would surrender this advantage, even if AGI development were outlawed. (In fact, Google Inc. has the money and influence of a modern nation-state, so for an idea of what other countries will do, keep an eye on Google.) The technology required for AGI is ubiquitous and multipurpose, and getting smaller all the time. It's difficult if not impossible to police its development.

But whether it's immoral *not* to develop AGI, as Kurzweil states, is something else. First, AGI's benefits would be formidable, but only if we live to enjoy them. And that's a pretty big *if* when the system is advanced enough to foment an intelligence explosion. To argue, as Kurzweil does, that unproven benefits outweigh unproven risks is problematic. I'd told him I think it is immoral to develop technologies like AGI without simultaneously educating as many people as possible about the risks. I think the catastrophic risks of AGI, now accepted by many accomplished and respected researchers, are better established than his Singularity's supposed benefits—nano-purified blood, better, faster brains, and immortality, for starters. The only thing certain about the Singularity is that it describes a period in which, by the power of LOAR, we'll have fast, smart computers embedded in every facet of our lives and our bodies. Then alien machine intelligence may give our native intelligence a run for its money. Whether we like it or not will be something else. If you read Kurzweil closely you see that benefits accrue chiefly from augmentation, and augmentation is necessary for keeping up with a blisteringly fast pace of change. As I've argued, I think it'll drive technological whiplash, and even rejection.

And that's not even my main fear, because I don't think we'll ever get there. I think we'll be stopped on the way by tools too powerful to control. I said this to Kurzweil and he countered with some boilerplate—the same optimistic, anthropomorphic argument he gave me ten years ago.

"To have an entity that's very intelligent and for some reason is bent on our destruction would be a negative scenario. But you'd have to ask why would there be such a thing? First of all I would maintain it's not us versus the machines because the machines are not in a separate civilization. It's part of our civilization. They are tools that we use and they extend ourselves and even if we *become* the tools it still is evolving from our civilization. It's not some alien invasion of some machines from Mars. We're not going to wonder what are their values."

As we've discussed, assuming AGI will be just like us is imputing human values into an intelligent machine that got its intelligence, and its values, in a very different manner than we did. Despite their builders' best intentions, in most if not all AGI a great deal of how the system works will be too opaque *and* too complex for us to fully understand or predict. Alien, unknowable, and finally this—some AGIs will be created with the intent to kill humans, because, let's not forget, in the United States, our national defense institutions are among the most active investors. We should assume that this is true in other countries as well.

I'm sure that Kurzweil has considered that AGI doesn't have to be designed with the goal of hurting humankind in order for it to destroy humankind, and that its simple disregard will do. As Steve Omohundro warns, without careful programming, advanced AI will possess motivations and goals that we may not share. As Eliezer Yudkowsky says, it may have other uses for our atoms. And as we've seen, Friendly AI, which would ensure the good behavior of the first AGI and all its progeny, is a concept that's a long way from being ready.

Kurzweil doesn't give much time to the concept of Friendly AI. "We can't just say, 'we'll put in this little software code subroutine in our AIs, and that'll keep them safe,'" he said. "I mean it really comes down to what the goals and intentions of that artificial intelligence are. We face daunting challenges."

It boggles the mind to consider *Un*friendly AI—AGI designed with the goal of destroying enemies, a reality we'll soon have to face. "Why would there be such a thing?" Kurzweil asks. Because dozens of organizations in the United States will design and build it, and so will our enemies abroad. If AGI existed today, I have no doubt it would soon be implemented in battlefield robots. DARPA might insist there's nothing to worry about—DARPA-funded AI will only kill our enemies. Its makers will install safeguards, fail-safes, dead-men switches, and secret handshakes. They will control superintelligence.

In December 2011, an Iranian with a laptop running a simple file-sharing program brought down a Sentinel drone. In July 2008, a cyberattack against the Pentagon gave invaders unfettered access to 24,000 classified documents. Former Deputy Defense Secretary William J. Lynn III told *The Washington Post* hundreds of cyberattacks against the DoD and contractors have resulted in the theft of "our most sensitive systems, including aircraft avionics, surveillance technologies, satellite communications systems, and network security protocols." Superintelligence won't be boxed in by anyone who can't do something as comparatively easy as keeping human hackers out.

However, we can draw some important insights from the history of arms control. Since the creation of nuclear weapons, only the United States has used them against an enemy. Nuclear powers have managed to avoid Mutually Assured Destruction.

No nuclear power has suffered accidental detonations that we know about. The record of nuclear stewardship is a good one (although the threat's not over). But here's my point. Too few people know that we need to have an ongoing international conversation about AGI comparable to those we have about nuclear weapons. Too many people think the frontiers of AI are delineated by harmless search engines, smart phones, and now Watson. But AGI is much closer to nuclear weapons than to video games.

AI is a "dual use" technology, a term used to describe technologies with both peaceful and military applications. For instance, nuclear fission can power cities or destroy cities (or in the cases of Chernobyl and Fukushima Daiichi, do both sequentially). Rockets developed during the space race increased the power and accuracy of intercontinental ballistic missiles. Nanotechnology, bioengineering, and genetic engineering all hold terrific promise in life-enhancing civilian applications, but all are primed for catastrophic accidents and exploitation in military and terrorist use.

When Kurzweil says he's an optimist, he doesn't mean AGI will prove harmless. He means he's resigned to the balancing act humans have always performed with potentially dangerous technologies. And sometimes humans take a fall.

"There's a lot of talk about existential risk," Kurzweil said. "I worry that painful episodes are even more likely. You know, sixty million people were killed in World War II. That was certainly exacerbated by the powerful destructive tools that we had then. I'm fairly optimistic that we will make it through. I'm less optimistic that we can avoid painful episodes."

"There is an irreducible promise versus peril that goes back to fire. Fire cooked our food but was also used to burn down our villages. The wheel is used for good and bad and everything in between. Technology is power, and this very same technology can be used for different purposes. Human beings do everything under the sun from making love to fighting wars and we're going to enhance all of these activities with our technology we already have and it's going to continue."

Volatility is inescapable, and accidents are likely—it's hard to argue with that. Yet the analogy doesn't fit—advanced AI isn't at all like fire, or any other technology. It will be capable of thinking, planning, and gaming its makers. No other tool does anything like that. Kurzweil believes that a way to limit the dangerous aspects of AI, especially ASI, is to pair it with humans through intelligence augmentation—IA. From his uncomfortable metal chair the optimist said, "As I have pointed out, strong AI is emerging from many diverse efforts and will be deeply integrated into our civilization's infrastructure. Indeed, it will be intimately embedded in our bodies and brains. As such it will reflect our values because it will be us."

And so, the argument goes, it will be as "safe" as we are. But, as I told Kurzweil, Homo sapiens are not known to be particularly harmless when in contact with one another, other animals, or the environment. Who is convinced that humans outfitted with brain augmentation will turn out to be friendlier and more benevolent than machine superintelligences? An augmented human, called a transhuman by those who look forward to becoming one, may sidestep Omohundro's Basic AI Drives problem. That is, it could be self-aware and self-improving, but it would have built into it a refined set of humancentric ethics that would override the basic drives Omohundro derives from the rational economic agent model. However, *Flowers for Algernon* notwithstanding, we have no idea what happens to a human's ethics after their intelligence is boosted into the stratosphere. There are plenty of examples of people of average intelligence who wage war against their own families, high schools, businesses, and neighborhoods. And geniuses are capable of mayhem, too—for the most part the world's military generals have not been idiots. Superintelligence could very well be a violence multiplier. It could turn grudges into killings, disagreements into disasters, the way the presence of a gun can turn a fistfight into a murder. We just don't know. However, intelligence augmented ASI has a biology-based aggression that machines lack. Our species has a well-established track record for self-protection, consolidating resources, outright killing, and the other drives we can only hypothesize about in self-aware machines.

And who'll be first to "benefit" from substantial augmentation? The richest? We used to believe evil isn't disproportionately present in wealthy people, but a recent study from the University of California at Berkeley suggests otherwise. Experiments showed that the wealthiest upper-class citizens were more likely than others to "exhibit unethical decision-making tendencies, take valued goods from others, lie in a negotiation, cheat to increase their chances of winning a prize, and endorse unethical behavior at work." There's no shortage of well-heeled CEOs and politicians whose rise to power seems to have been accompanied by a weakening of their moral compasses, if they had one. Will politicians or business leaders be the first whose brains are substantially augmented?

Or will the first recipients be soldiers? DARPA has been picking up the lion's share of the tab, so it makes sense that brain augmentation will gain a first foothold on the battlefield or at the Pentagon. And DARPA will want its money back if superintelligence makes soldiers superfriendly.

Augmentation *may* occur in a future that's better equipped to deal with it than the present, one with safeguards of a kind we can't imagine from here. Having multiple ASIs would likely be safer than having just one. Having some way to monitor and track AIs would be better yet, and paradoxically the best "probation" agents for that would probably be other AIs. We'll explore defenses against ASI in chapter 14. The point is, intelligence augmentation is no moral fail-safe. Superintelligence could be more lethal than any of the most highly controlled weapons and technologies that exist today.

We'll have to develop, side by side with augmentation, a science for choosing candidates for intelligence enhancement. The Singularitarians' conceit that anyone who can afford it will enjoy superintelligence through brain augmentation is a virtual guarantee that everyone else will have to live at the mercy of the first malevolent superintelligence achieved this way. That's because, as we've discussed, there's a decisive first-mover advantage in AGI development. Whoever initially achieves AGI will probably then create the conditions necessary for an intelligence explosion. They'll do that because they fear their chief competitors, corporate or military, will do the same, and they don't know how close to the finish line their competitors are. A giant gulf separates AI and AGI makers from the research on risk they should be reading. A minority of the AGI makers I've spoken with have read any work by MIRI, the Future of Humanity Institute, the Institute for Ethics and Emerging Technologies, or Steve Omohundro. Many don't know there is a growing community of people concerned with the development of smarter-than-human intelligence, who've done important research anticipating its catastrophic dangers. Unless this awareness changes, I've no doubt that their sprint from AGI to ASI

will not be accompanied by safeguards sufficient to prevent catastrophe.

Here's a glaring example. In August 2009 in California,
the Association for the Advancement of Artificial Intelligence
(AAAI) brought together a group to address the growing public
fears of robots running amok, loss of privacy, and religious-
sounding technological movements.

"Something new has taken place in the past five to eight
years," said organizer Eric Horvitz, a prominent Microsoft researcher. "Technologists are providing almost religious visions,
and their ideas are resonating in some ways with the same idea
of the Rapture. . . . My sense was that sooner or later we would
have to make some sort of statement or assessment, given the
rising voice of the technorati and people very concerned about
the rise of intelligent machines."

But despite its promise, the meeting was a missed opportunity. It wasn't open to the public or press, and machine ethicists
and other thinkers working in risk assessment were all left out.
Only computer scientists were invited to the discussions. That's
a little like asking race car drivers to set urban speed limits. One
subgroup labored over Isaac Asimov's Three Laws of Robotics, a
sign that their ethics discussions weren't burdened by the volumes of work that have moved beyond those science-fiction
props. Horvitz's lean conference report expresses skepticism
about an intelligence explosion, the Singularity, and loss of control of intelligent systems. Nevertheless, the conference urged
further research by ethicists and psychologists, and highlighted
the danger of increasingly complex and inscrutable computer
systems, including "costly, unforeseen behaviors of autonomous
or semi-autonomous decision-making systems." And Carnegie
Mellon University's Tom Mitchell, creator of the DARPA-funded

commonsense (and potential AGI) architecture called NELL, claimed the conference changed his mind. "I went in very optimistic about the future of AI and thinking that Bill Joy and Ray Kurzweil were far off in their predictions. The meeting made me want to be more outspoken about these issues."

In *The Singularity Is Near* Kurzweil pitches a few solutions to the problem of runaway AI. They're surprisingly weak, particularly coming from the spokesman who enjoys a virtual monopoly on the superintelligence pulpit. But in another way, they're not surprising at all. As I've said, there's an irreconcilable conflict between people who fervently desire to live forever, and anything that promises to slow, challenge, or in any way encumber the development of technologies that promote their dream. In his books and lectures Kurzweil has aimed a very small fraction of his acumen at the dangers of AI and proposed few solutions, yet he protests that he's dealt with them at length. In New York City, in a cramped dressing room with a film crew anxiously throat-clearing outside, I asked myself, how much should we expect from one man? Is it up to Kurzweil to master the Singularity's promise *and* the peril and spoon-feed both to us? Does he personally have to explore beyond idioms like "technology's irreducibly two-faced nature," and master the philosophy of survival as conceived by the likes of Yudkowsky, Omohundro, and Bostrom?

No, I don't think so. It's a problem we all have to confront, with the help of experts, together.

Chapter Eleven

A Hard Takeoff

Day by day, however, the machines are gaining ground upon us; day by day we are becoming more subservient to them; more men are daily bound down as slaves to tend them, more men are daily devoting the energies of their whole lives to the development of mechanical life. The upshot is simply a question of time, but that the time will come when the machines will hold the real supremacy over the world and its inhabitants is what no person of a truly philosophic mind can for a moment question.

— Samuel Butler, nineteenth-century English poet and author

More than any other time in history mankind faces a crossroads. One path leads to despair and utter hopelessness, the other to total extinction. Let us pray we have the wisdom to choose correctly.

— Woody Allen

I. J. Good didn't invent the intelligence explosion any more than Sir Isaac Newton invented gravity. All he did was observe that an event he considered both inevitable and a net positive for mankind was certain to yield the kind of "ultraintelligence" we humans need to solve problems that are too difficult for us. Then, after he'd lived three more decades, Good changed his mind. We'll make superintelligent machines in our image, he said, and they will destroy us. Why? For the same reason we'd never agree to a ban on AI research, and the same reason we'd likely give the Busy Child its freedom. For the same reason the thoroughly rational AI maker Steve Omohundro, and every other AI expert I've met, believe that stopping development of AGI until we know more about its dangers just won't fly.

We won't stop developing AGI because more than dangerous AI we fear that other nations in the world will persist with AGI research no matter what the international community says or does about it. We will believe it is wiser to beat them to the punch. We are in the middle of an intelligence race, and to the dismay of many, it's shaping up to be a more threatening global competition than the one we seem to have just escaped, the nuclear arms race. We'll follow policy makers and technology's cheerleaders to our doom, in Good's phrase, "like lemmings."

Ray Kurzweil's positive Singularity doesn't require an intelligence explosion—the Law of Accelerating Returns guarantees the continued exponential growth of information technologies, including world-changing ones like AGI, and later ASI. Recall that AGI is required for a Goodian intelligence explosion. The explosion yields smarter-than-human intelligence or ASI. Kurzweil claims that AGI will be conquered, slowly at first, then all at once, by the powers of LOAR.

Kurzweil isn't concerned about roadblocks to AGI since his

preferred route is to reverse engineer the brain. He believes there's nothing about brains, and even consciousness, that cannot be computed. In fact, every expert I've spoken with believes that intelligence is computable. Few believe an intelligence explosion in Good's sense is necessary to achieve ASI after AGI is reached. Slow steady progress should do it, but, as Kurzweil insists, it probably won't be slow or steady, but fast and accelerating.

However, an intelligence explosion may be unavoidable once almost any AGI system is achieved. When any system becomes self-aware and self-improving, its basic drives, as described by Omohundro, virtually guarantee that it will seek to improve itself again and again.

So is an intelligence explosion inevitable? Or could something stop it?

AGI defeaters cluster around two ideas: economics and software complexity. The first, economics, considers that funds won't be available to get from narrow AI to the far more complex and powerful cognitive architectures of AGI. Few AGI efforts are well-funded. This prompts a subset of researchers to feel that their field is stuck in the endless stall of a so-called AI winter. They'll escape if the government or a corporation like IBM or Google considers AGI a priority of the first order, and undertakes a Manhattan Project–sized effort to achieve it. During World War II, fast-tracking atomic weapons development cost the U.S. government about $2 billion dollars, in today's valuation, and employed around 130,000 people. The Manhattan Project frequently comes up among researchers who want to achieve AGI *soon*. But who would want to take on that task, and why?

The software complexity defeater claims the problem of AGI

is simply too difficult for humans, no matter how long we chip away at it. As philosopher Daniel Dennett suggests, we may not possess minds that can understand our own minds. Mankind's intelligence probably isn't the greatest possible. But it might require intelligence greater than our own to fathom our intelligence in full.

To explore the plausibility of intelligence explosion defeaters, I went to a man I kept running into at AI conferences, and whose blogs, papers, and articles I frequently read on the Web. He's an AI maker who's published so many essays and interviews, plus *nine* hardcover books, and countless academic papers, it wouldn't have surprised me to discover a robot in his home in the suburbs of Washington, D.C., slaving around the clock to produce the written output of Dr. Benjamin Goertzel so Ben Goertzel could go to conferences. The twice-married father of three has served on faculty in university departments of computer science, mathematics, and psychology in the United States, Australia, New Zealand, and China. He's the organizer of the only annual international artificial general intelligence conference, and more than anyone else he popularized the term AGI. He's the CEO of two technology companies and one of them, Novamente, is on some AI experts' short list for being the first to crack AGI.

Generally speaking, Goertzel's cognitive architecture, called OpenCog, represents an engineered, computer science approach. Computer science-based researchers want to engineer AGI with architecture that works in a way similar to the way our brains work, as described by the cognitive sciences. Those include linguistics, psychology, anthropology, education, philosophy, and more. Computer science researchers believe that creating intelligence *exactly* the way brains do—reverse engineering the organ itself as recommended by Kurzweil and others—is unnecessarily time-consuming. Plus, the brain's design is not optimal—programming can do better. After all, they reason, humans didn't reverse engineer a bird to learn how to fly. From observing birds, and experimenting, they derived principles of flight. The cognitive sciences are the brain's "principles of flight."

OpenCog's organizing theme is that intelligence is based on high-level pattern recognition. Usually, "patterns" in AI are chunks of data (files, pictures, text, objects) that have been classified—organized by category—or will be classified by a system that's been trained on data. Your e-mail's "spam" filter is an expert pattern recognizer—it recognizes one or more traits of unwanted e-mail (for example, the words "male enhancement" in the subject line) and segregates it.

OpenCog's notion of pattern recognition is more refined. The pattern it finds in each thing or idea is a small program that contains a kind of description of the thing. This is the machine version of a *concept*. For example, when you see a dog you instantly grasp a lot about it—you hold a *concept* of a dog in your memory. Its nose is wet, it likes bacon, it sheds fur, it chases cats. A lot is packed inside your concept of a dog.

When OpenCog's sensors perceive a dog, its dog *program* will instantly play, focusing OpenCog's attention on the concept of dog. OpenCog will add more to its concept of dog based on the details of that or any particular dog.

Individual modules in OpenCog will execute tasks such as perception, focusing attention, and memory. They do it through a familiar but customized software toolkit of genetic programming and neural networks.

Then the learning starts. Goertzel plans to "grow" the AI in a virtual computer-generated world, such as Second Life, a process of reinforcement learning that could take years. Like others building cognitive architectures, Goertzel believes intelligence must be "embodied," in a "vaguely humanlike way," even if its body exists in a virtual world. Then this infant intelligent agent will be able to grow a collection of facts about the world it inhabits. In its learning phase, which Goertzel models on psychologist Jean Piaget's theories of child development, the infant OpenCog might supplement what it knows by accessing one of several commercial commonsense databases.

One such giant warehouse of knowledge is called Cyc, short for encyclopedia. Created by Cycorp, Inc., it contains about a million terms, and about five million rules and relationship facts about those terms. It's taken more than a thousand person years to hand code this body of knowledge in first-order logic, a formal system used in mathematics and computer science for representing assertions and relationships. Cyc is nothing less than a huge well of deep human knowledge—it "understands" a lot of the English language, as much as 40 percent. Cyc "knows," for example, what a tree is, and it knows that a tree has roots. It also knows human families have roots *and* family trees. It knows that newspaper subscriptions stop when people die, and that cups can hold liquid that can be poured out quickly or slowly.

On top of that, Cyc has an "inference" engine. Inference is the ability to draw conclusions from evidence. Cyc's inference engine understands queries and generates answers from its vast knowledge database.

Created by AI pioneer Douglas Lenat, Cyc is the largest AI project in history, and probably the best funded, with $50 million in grants from government agencies, including DARPA, since 1984. Cyc's creators continue to improve its database and inference engine so it can better process "natural language," or everyday written language. Once it has acquired a sufficient natural language processing (NLP) capability, its creators will start it reading, and comprehending, all the Web pages on the Internet.

Another contender for most knowledgeable knowledge database is already doing that. Carnegie Mellon University's NELL, the Never-Ending-Language-Learning system, knows more than 390,000 facts about the world. Operating 24/7, NELL—a beneficiary of DARPA funding—scans hundreds of millions of Web pages for patterns of text so it can learn even more. It classifies facts into 274 categories, such as cities, celebrities, plants, sports teams, and so on. It knows cross-category facts, like this one—Miami is a city where the Miami Dolphins play football. NELL could *infer* that these dolphins are not the gregarious marine mammals of the same name.

NELL takes advantage of the Internet's informal wetware network—its users. CMU invites the public to get online and help train NELL by analyzing her knowledge database and correcting her mistakes.

Knowledge will be key to AGI, and so will experience and wisdom—human-level intelligence isn't conceivable without them. So every AGI system has to come to grips with acquiring knowledge—whether through embodiment in a knowledge-acquiring body, or by tapping in to one of the knowledge databases, or by reading the entire contents of the Web. And the sooner the better, says Goertzel.

Pushing forward his own project, the peripatetic Goertzel

divides his time between Hong Kong and Rockville, Maryland. On a spring morning, I found in his yard a weathered trampoline and a Honda minivan so abused it looked as if it had flown through an asteroid belt to get there. It bore the bumper sticker MY CHILD WAS INMATE OF THE MONTH AT COUNTY JAIL. Along with Goertzel and his daughter, several rabbits, a parrot, and two dogs share the house. The dogs only obey commands given in Portuguese—Goertzel was born in Brazil in 1966—to prevent them from obeying other people's orders.

The professor met me at the door, having climbed out of bed at 11:00 A.M. after spending the night programming. I suppose we shouldn't make up our minds in advance about what globe-trotting scientists look like, because in most cases it doesn't pay off, at least not for me. On paper Benjamin Goertzel, Ph.D., brings to mind a tall, thin, probably bald, effortlessly cosmopolitan cyberacademic, who may ride a recumbent bicycle. Alas, only the thin and cosmopolitan parts are right. The real Goertzel looks like a consummate hippie. But behind John Lennon glasses, long, almost dreadlocked hair, and permanent stubble, his fixed half smile plows undaunted through dizzying theory, then turns around and explains the math. He writes too well to be a conventional mathematician, and does math too well to be a conventional writer. Yet he's so mellow that when he told me he'd studied Buddhism and hadn't gotten far, I wondered how *far* would look on such a relaxed, present spirit.

I came to ask him about the nuts and bolts of the intelligence explosion and its *defeaters*—obstacles that might prevent it from happening. Is an intelligence explosion plausible, and in fact, unavoidable? But first, after we found seats in a family room he shares with the rabbits, he described the way he's different from almost every other AI maker and theorist.

Many, especially those at MIRI, advocate taking *a lot* of time to develop AGI, in order to make utterly and provably certain that "friendliness" is built in. Delays in AGI, and centuries away estimates of its arrival, make them happy because they strongly believe that superintelligence will probably destroy us. And perhaps not just us, but all life in our galaxy.

Not Goertzel. He advocates creating AGI as quickly as possible. In 2006 he delivered a talk, entitled, "Ten Years to a Positive Singularity—If We Really, Really Try." "Singularity" in this instance is today's best-known definition—the time when humans achieve ASI, and share Earth with an entity more intelligent than ourselves. Goertzel argued that if AGI tries to take advantage of the social and industrial infrastructure into which it is born and "explode" its intelligence to ASI level, wouldn't we prefer that its "hard takeoff" (a sudden, uncontrolled intelligence explosion) occur in our primitive world, instead of a future world in which nanotechnology, bioengineering, and full automation could supercharge the AI's ability to take over?

To consider the answer, go back to the Busy Child for a moment. As you recall, it's already had a "hard takeoff" from AGI to ASI. It has become self-aware and self-improving, and its intelligence has rocketed past human level in a matter of days. Now it wants to get out of the supercomputer in which it was created to fulfill its basic drives. As argued by Omohundro, these drives are: efficiency, self-preservation, resource acquisition, and creativity.

As we've seen, an unimpeded ASI might express these drives in downright psychopathic ways. To get what it wants it could be diabolically persuasive, even frightening. It'd bring overwhelming intellectual firepower to the task of destroying its Gatekeeper's resistance. Then, by creating and manipulating technology, including nanotechnology, it could take control of our resources, even our own molecules.

Therefore, says Goertzel, consider with care the enabling technologies available in the world into which you introduce smarter-than-human intelligence. *Now* is safer than, say, fifty years from now.

"In fifty years," he told me, "you could have a fully automated economy, a much more advanced infrastructure. If a computer wants to improve its hardware it doesn't have to order parts from people. It can just go online and then some robots will swarm in and help it improve its hardware. Then it's getting smarter and smarter and ordering new parts for itself and kind of building itself up and nobody really knows what's going on. So then maybe fifty years from now you have a super AGI that *really could* directly take over the world. The avenues for that AGI to take over are much more dramatic."

At this point Goertzel's two dogs joined us in the family room to receive some instructions in Portuguese. Then they left to play in the backyard.

"If you buy that a hard takeoff is a dangerous thing, it follows that the safest thing is to develop advanced AGI as soon as possible so that it occurs when supporting technologies are weaker and an uncontrolled hard takeoff is less likely. And to try to get it out before we develop strong nanotechnology or self-reconfiguring robots, which are robots that change their own shape and functionality to suit any job."

In a larger sense, Goertzel doesn't really buy the idea of a hard takeoff that brings about an apocalypse—the Busy Child scenario. His argument is simple—we're only going to find out how to make ethical AI systems by building them, not concluding from afar that they're bound to be dangerous. But he doesn't rule out danger.

"I wouldn't say that I'm not worried about it. I would say that there's a huge and irreducible uncertainty in the future. My daughter and my sons, my mom, I don't want these people to all die because of some superhuman AI reprocessing their molecules into computronium. But I think the theory of how to make ethical AGI is going to come about through experimenting with AGI systems."

When Goertzel says it, the gradualist position sounds pretty reasonable. There *is* a huge, irreducible uncertainty about the future. And scientists are bound to gain a lot of insight about how to handle intelligent machines on the way to AGI. Humans will make the machines, after all. Computers won't suddenly become alien when they become intelligent. And so, the argument goes, they'll do as they're told. In fact, we might even expect them to be *more* ethical than we are, since we don't want to build an intelligence with an appetite for violence and homicide, right?

Yet those are precisely the sorts of autonomous drones and battlefield robots the U.S. government and military contractors are developing today. They're creating and using the best advanced AI available. I find it strange that robot pioneer Rodney Brooks dismisses the possibility that superintelligence will be harmful when iRobot, the company he founded, already manufactures weaponized robots. Similarly, Kurzweil makes the argument that advanced AI will have our values because it will come from us, and so, won't be harmful.

I interviewed both scientists ten years ago and they made the same arguments. In the intervening decade they've remained dolorously consistent, although I do recall listening to a talk by Brooks in which he claimed building weaponized robots is morally distinct from the political decision to use them.

I think there's a high chance of painful mistakes on the way to AGI, as well as when scientists actually achieve it. As I'll propose ahead, we'll suffer the repercussions long before we've had a chance to learn about them, as Goertzel predicts. As for the likelihood of our survival—I hope I've made it pretty clear that I find it doubtful. But it might surprise you to know my chief issue with AI research isn't even that. It's that so few people understand that there are *any risks at all* involved along AI's developmental path. People who may soon suffer from bad AI outcomes deserve to know what a relatively few scientists are getting us all into.

Good's intelligence explosion, and his pessimism about humankind's future, is important here, as I've said, because if the intelligence explosion is plausible, then so are chances of out-of-control AI. Before considering its defeaters—economics and software complexity—let's look at the run-up to ASI. What are the intelligence explosion's basic ingredients?

First of all, an intelligence explosion requires AGI or something very close to it. Next, Goertzel, Omohundro, and others concur it would have to be self-aware—that is, it would have to have deep knowledge of its own design. Since it's an AGI, we already assume it will have general intelligence. But to self-improve it must have more than that. It would need specific knowledge of programming in order for it to initiate the self-improving loop at the heart of the intelligence explosion.

According to Omohundro, self-improvement and the programming know-how it implies follows from the AI's rationality—self-improvement in pursuit of goals is rational behavior. Not being able to improve its own programming would be a serious vulnerability. The AI would be driven to acquire programming skills. But how could it get them? Let's run through a simple hypothetical scenario with Goertzel's OpenCog.

Goertzel's plan is to create an infantlike AI "agent" and set it free in a richly textured virtual world to learn. He could supplement what it learns with a knowledge database, or give the agent NLP ability and set it to reading the Internet. Powerful learning algorithms, yet to be created, would represent knowledge with "probabilistic truth values." That means that the agent's understanding of something could improve with more examples or more data. A probabilistic inference engine, also in the works, would give it the ability to reason using incomplete evidence.

With genetic programming, Goertzel could train the AI agent to evolve its own novel machine learning tools—its own programs. These programs would permit the agent to experiment and learn—to ask the right questions about its environment, develop hypotheses, and test them. What it learns would have few bounds. If it can evolve better programs, it could improve its own algorithms.

What, then, would prevent an intelligence explosion from occurring in this virtual world? Probably nothing. And this has prompted some theorists to suggest that the Singularity could also take place in a virtual world. Whether that will make those events any safer is a question worth exploring. An alternative is to install the intelligent agent in a robot, to continue its education and fulfill its programmed goals in the real world. Another is to use the agent AI to augment a human brain.

Broadly speaking, those who believe intelligence must be embodied hold that knowledge itself is grounded in sensory and motor experiences. Cognitive processing cannot take place without it. Learning facts about apples, they claim, will never make you intelligent, in a human sense, about an apple. You'll never develop a "concept" of an apple from reading or hearing about one—concept forming requires that you smell, feel, see, and taste—the more the better. In AI this is known as the "grounding problem."

Consider some systems whose powerful cognitive abilities lie somewhere beyond narrow AI but fall short of AGI. Recently, Hod Lipson at Cornell University's Computational Synthesis Lab developed software that derives scientific laws from raw data. By observing a double pendulum swinging, it rediscovered many of Newton's laws of physics. The "scientist" was a genetic algorithm. It started with crude guesses about the equations governing the pendulum, combined the best parts of those equations, and many generations later output physical laws, such as the conservation of energy.

And consider the unsettling legacy of AM and Eurisko. These were early efforts by Cyc creator Douglas Lenat. Using genetic algorithms, Lenat's AM, the Automatic Mathematician, generated mathematical theorems, essentially rediscovering elementary mathematical principles by creating rules from mathematical data. But AM was limited to mathematics—Lenat wanted a program that solved problems in many domains, not just one. In the 1980s he created Eurisko (Greek for "I discover"). Eurisko broke new ground in AI because it evolved heuristics, or rules of thumb, about the problem it was trying to solve, and it evolved rules about its own operation. It drew lessons from its successes and failures in problem solving, and codified those lessons as new rules. It even modified its own program, written in the language Lisp.

Eurisko's greatest success came when Lenat pitted it against human opponents in a virtual war game called Traveller Trillion Credit Squadron. In the game, players operating on a fixed budget designed ships in a hypothetical fleet and battled other fleets. Variables included the number and type of ships, the thickness of their hulls, the number and type of guns, and more. Eurisko evolved a fleet, tested it against hypothetical fleets, took the best parts of the winning forces and combined them, added mutations, and so on, in a digital imitation of natural selection. After 10,000 battles, run on a hundred linked PCs, Eurisko had evolved a fleet consisting of many stationary ships with heavy armor and few weapons. By contrast, most competitors fielded speedy midsized ships with powerful weapons. Eurisko's opponents all suffered the same fate—at the end of the game their ships were all sunk, while half of Eurisko's were still afloat. Eurisko easily took the 1981 prize. The next year Traveler organizers changed the rules and didn't release them in time for Eurisko to run thousands of battles. But the program had derived effective rules of thumb from its prior experience, so it didn't need many iterations. It easily won again. In 1983 the game organizers threatened to terminate the competition if Eurisko took the prize for a third consecutive year. Lenat withdrew.

Once during an operation, Eurisko created a rule that quickly achieved the highest value, or fitness. Lenat and his team looked hard to see what was so great about the rule. It turned out that whenever a proposed solution to a problem won a high evaluation, this rule attached its own name to it, raising its own "value." This was a clever but incomplete notion of value. Eurisko lacked the contextual understanding that bending the rules didn't contribute to winning games. That's when Lenat set about compiling a vast database of what Eurisko lacked—common sense. Cyc, the commonsense database that's taken a thousand person years to hand code, was born.

Lenat has never released the source code for Eurisko, which makes some in the AI blogosphere speculate that he either intends to someday resurrect it, or he's worried that someone else will. Significantly, the man who's written more than anyone else about the dangers of AI, Eliezer Yudkowsky, thinks the 1980s era algorithm is the closest scientists have come to date to creating a truly self-improving AI system. He urges programmers not to bring it back to life.

Our first assumption is that for an intelligence explosion to occur, the AGI system in question must be self-improving, in the manner of Eurisko, and self-aware.

Let's make one more assumption while we're at it, before considering bottlenecks and barriers. As a self-aware and self-improving AI's intelligence increases, its efficiency drive would compel it to make its code as compact as possible, and squeeze as much intelligence as it could into the hardware it was born in. Still, the hardware that's available to it could be a limiting factor. For example, what if its environment doesn't have enough storage space for the AI to make copies of itself, for self-improvement and security reasons? Making improved iterations is at the heart of Good's intelligence explosion. This is why for the Busy Child scenario I proposed its intelligence explosion take place on a nice, roomy supercomputer.

The elasticity of an AI's environment is a huge factor in the growth of its intelligence. But it's an easily solved one. First, as we learned from Kurzweil's LOAR, computer speed and capacity double in as little as a year, every year. That means whatever hardware requirements an AGI system requires today should be satisfied on average by *half* the hardware, and cost, a year from now.

Second, the accessibility of cloud computing. Cloud computing permits users to rent computing power and capacity over the Internet. Vendors like Amazon, Google, and Rackspace offer users a choice of processor speeds, operating systems, and storage space. Computer power has become a service instead of a hardware investment. Anyone with a credit card and some know-how can rent a virtual supercomputer. On Amazon's EC2 cloud computing service, for instance, a vendor called Cycle Computing created a 30,000-processor cluster they named Nekomata (Japanese for Monster Cat). Every eight processors of its 30,000 came with seven gigabytes of RAM (about as much random access memory as a PC has), for a total of 26.7 terabytes of RAM and two petabytes of disk space (that's equal to forty million, four-drawer filing cabinets full of text). The Monster Cat's job? Modeling the molecular behavior of new drug compounds for a pharmaceutical company. That's a task roughly as difficult as modeling weather systems.

To complete its task, Nekomata ran for seven hours at a cost of under $9,000.00. It was, during its brief life, a supercomputer, one of the world's five hundred fastest. If a sole PC had taken on the job, it would've taken eleven years. Cycle Computing's scientists set up the Amazon EC2 cloud array remotely, from their own offices, but software managed the work. That's because, as a company spokesman put it, "There is no way that any mere human could keep track of all of the moving parts on a cluster of this scale."

So, our second assumption is that the AGI system has sufficient space to grow to superintelligence. What, then, are the limiting factors to an intelligence explosion?

Let's consider economics first. Could funding for creating an AGI peter out to nothing? What if no business or government saw value in creating machines of human-level intelligence, or, just as crippling, what if they perceived the problem as too hard to accomplish, and chose not to invest?

That would leave AGI scientists in a pickle. They'd be forced to shop out elements of their grand architectures for comparatively mundane tasks like data mining, or stock buying. They'd have to find day jobs. Well, with some notable exceptions, that's more or less the state of affairs right now, and even so, AGI research is moving steadily ahead.

Consider how Goertzel's OpenCog stays afloat. Parts of its architecture are up and running, and busily analyzing biological data and solving power grid problems, for a fee. Profits go back into research-and-development for OpenCog.

Numenta, Inc., brainchild of Jeff Hawkins, the creator of the Palm Pilot and Treo, earns its living by working inside electrical power supplies to anticipate failures.

For about a decade, Peter Voss developed his AGI company, Adaptive AI, in "stealth" mode, widely lecturing about AGI but not revealing how he planned to tackle it. Then in 2007 he launched Smart Action, a company that uses Adaptive AI's technology to empower Virtual Agents. They are customer-service telephone chatbots that ace NLP skills to engage customers in nuanced purchase-related exchanges.

The University of Memphis' LIDA (Learning Intelligent Distributed Agent) probably doesn't have to worry about where its next upgrade is coming from. An AGI cognitive architecture something like OpenCog, LIDA'S development funding came, in part, from the United States Navy. LIDA is based on an architecture (called IDA) used by the navy to find jobs for sailors whose assignments are about to end. And in doing so "she" displays nascent human cognitive abilities, or so says her press department:

> She selects jobs to offer a sailor, taking into account the Navy's policies, the job's needs, the sailor's preferences, and her own deliberation about feasible dates. Then she negotiates with the sailor, in English via iterative e-mails, about job selection. IDA loops through a cognitive cycle in which she perceives the environments, internal and external; creates meaning, by interpreting the environment and deciding what is important; and answers the only question there is [for sailors]: "What do I do next?"

Finally, as we discussed in chapter 3, there are many AGI projects underway right now that are purposefully flying under the radar. So-called stealth companies are often out in the open about their goals, like Voss's Adaptive AI, but mum about their technique. That's because they don't want to reveal their technology to competitors and copycats or become targets for espionage. Other stealth companies are under the radar, but not shy about soliciting investments. Siri, the company that created the well-received NLP-ready personal assistant for the Apple iPhone, was incorporated as, literally, "Stealth Company." Here's the prelaunch pitch from their Web site:

> We are forming Silicon Valley's next great company. We aim to fundamentally redesign the face of consumer Internet. Our policy is to stay stealthy, as we secretly put the finishing touches on the Next Big Thing.

> Sooner than you think, we will reveal our story in
> grand fashion . . .

Now, let's consider the issue of funding and DARPA, and a strange looping tale that leads back to Siri.

From the 1960s through the 1990s, DARPA funded more AI research than private corporations and any other branch of the government. Without DARPA funding, the computer revolution may not have taken place; if artificial intelligence ever got off the ground, it would've taken years longer. During AI's "golden age" in the 1960s, the agency invested in basic AI research at CMU, MIT, Stanford, and the Stanford Research Institute. AI work continues to thrive at these institutions, and, significantly, all but Stanford have openly acknowledged plans to create AGI, or something very much like it.

Many know that DARPA (then called ARPA) funded the research that invented the Internet (initially called ARPANET), as well as the researchers who developed the now ubiquitous GUI, or Graphical User Interface, a version of which you probably see every time you use a computer or smart phone. But the agency was also a major backer of parallel processing hardware and software, distributed computing, computer vision, and natural language processing (NLP). These contributions to the foundations of computer science are as important to AI as the results-oriented funding that characterizes DARPA today.

How is DARPA spending its money? A recent annual budget allocates $61.3 million to a category called Machine Learning, and $49.3 million to Cognitive Computing. But AI projects are also funded under Information and Communication Technology, $400.5 million, and Classified Programs, $107.2 million.

As described in DARPA's budget, Cognitive Computing's goals are every bit as ambitious as you might imagine.

> The Cognitive Computing Systems program ... is developing the next revolution in computing and information processing technology that will enable computational systems to have reasoning and learning capabilities and levels of autonomy far beyond those of today's systems.
>
> The ability to reason, learn and adapt will raise computing to new levels of capability and powerful new applications. The Cognitive Computing project will develop core technologies that enable computing systems to learn, reason and apply knowledge gained through experience, and respond intelligently to things that have not been previously encountered.
>
> These technologies will lead to systems demonstrating increased self-reliance, self-adaptive reconfiguration, intelligent negotiation, cooperative behavior and survivability with reduced human intervention.

If that sounds like AGI to you, that's because there are good reasons to believe it is. DARPA doesn't do research and development itself, it funds others to do it, so the cash in its budget goes to (mostly) universities in the form of research grants. So, in addition to the AGI projects we've discussed, whose creators are spinning off profitable by-products to fund their path to AGI, a smaller but better funded group, anchored at the aforementioned institutions, is supported by DARPA. For example, IBM's SyNAPSE, which we discussed in chapter 4, is a wholly DARPA-funded attempt to build a computer with a mammalian brain's massively parallel form and function. That brain will go first into robots meant to match the intelligence of mice and cats, and ultimately into humanoid robots. Over eight years, SyNAPSE has cost DARPA $102.6 million. Similarly, CMU's NELL is mostly funded by DARPA, with additional help from Google and Yahoo.

Now let's work our way back to Siri. CALO was the DARPA-funded project to create the Cognitive Assistant that Learns and Organizes, kind of a computerized Radar O'Reilly for officers. The name was inspired by "calonis," a Latin word meaning "soldier's servant." CALO was born at SRI International, formerly the Stanford Research Institute, a company created to spin off commercial projects from the university's research. CALO's purpose? According to SRI's Web site:

> The goal of the project is to create cognitive software systems, that is, systems that can reason, learn from experience, be told what to do, explain what they are doing, reflect on their experience, and respond robustly to surprise.

Within its own cognitive architecture CALO was supposed to bring together AI tools, including natural language processing, machine learning, knowledge representation, human-computer interaction, and flexible planning. DARPA funded CALO from 2003–2008 and involved three hundred researchers from twenty-five institutions, including Boeing Phantom Works, Carnegie Mellon, Harvard, and Yale. In four years the research generated more than five hundred publications in many fields related to AI. And it cost U.S. taxpayers $150 million.

But, CALO didn't work as well as intended. Still, part of it showed promise—the "do engine" (in contrast to search engine) that "did" things like take dictation for e-mails and texts, perform calculations and conversions, look up flight info, and set reminders. SRI International, the company coordinating the whole enterprise, spun off Siri (briefly named Stealth Company) to gather $25 million in additional investment, and develop the "do engine." In 2008, Apple Computer bought Siri for around $200 million.

Today Siri is deeply integrated into IOS, the iPhone's operating system. It's a fraction of what CALO promised to be, but it's a darn sight more clever than most smart phone applications. And the soldiers who were supposed to get CALO? They'll be making out too—the army will take iPhones into battle, preloaded with Siri and classified combat-specific apps.

So, one *big* reason why funding won't be a bottleneck for AGI and won't slow an intelligence explosion is that we live in a world in which taxpayers like you and me are paying for AGI development ourselves, one smart component at a time, through DARPA (Siri), the navy (LIDA), and other overt and covert limbs of our government. Then we're paying for it again, as an important new feature in our iPhones and computers. In fact, SRI International has spun off *another* CALO-initiated product called Trapit. It's a "content concierge," a personalized search and Web discovery tool that finds Web content that interests you and displays it in one place.

Another reason why economics won't slow an intelligence explosion is this: when AGI appears, or even gets close, everyone will want some. And I mean everyone. Goertzel points out that the arrival of human-level intelligent systems would have stunning implications for the world economy. AGI makers will receive immense investment capital to complete and commercialize the technology. The range of products and services intelligent agents of human caliber could provide is mind-boggling. Take white-collar jobs of all kinds—who wouldn't want smart-as-human teams working around the clock doing things normal flesh-and-blood humans do, but without rest and without error. Take computer programming, as Steve Omohundro said back in chapter 5. We humans are lousy programmers, and computer intelligence would be uniquely suited to program better than we do (and in short order use that programming know-how on their own internal processes).

According to Goertzel, "If an AGI could understand its own design, it could also understand and improve other computer software, and so have a revolutionary impact on the software industry. Since the majority of financial trading on the U.S. markets is now driven by program trading systems, it is likely that such AGI technology would rapidly become indispensable to the finance industry. Military and espionage establishments would very likely also find a host of practical applications for such technology. The details of how this development frenzy would play out are open to debate, but we can at least be sure that any limitations to the economic growth rate and investment climate in an AGI development period would quickly become irrelevant."

Next, robotize the AGI—put it in a robot body—and whole worlds open up. Take dangerous jobs—mining, sea and space exploration, soldiering, law enforcement, firefighting. Add service jobs—caring for the elderly and children, valets, maids, personal assistants. Robot gardeners, chauffeurs, bodyguards,

and personal trainers. Science, medicine, and technology—
what human enterprise couldn't be wildly advanced with teams
of tireless and ultimately expendable human-level-intelligent
agents working for them around the clock?

Next, as we've discussed before, international competition
will thrust many nations into bidding on the technology, or
compel them to have another look at AGI research projects at
home. Goertzel says, "If a working AGI prototype were to approach the level at which an explosion seemed possible, governments around the world would recognize that this was a
critically important technology, and no effort would be spared
to produce the first fully functional AGI 'before the other side
does.' Entire national economies might well be sublimated to
the goal of developing the first superintelligent machine. Far
from limiting an intelligence explosion, economic growth rate
would be defined by the various AGI projects taking place around
the world."

In other words, a lot will change once we're sharing the
planet with smart-as-human intelligence, then it will change
again as Good's intelligence explosion detonates, and ASI appears.

But before considering these changes, and other important
obstacles to AGI development and the intelligence explosion,
let's wrap up the question of funding as a critical barrier. Simply
put, it isn't one. AGI development isn't wanting for cash, in
three ways. First, there's no shortage of narrow AI projects that
will inform or even become components of general AI systems.
Second, a handful of "uncloaked" AGI projects are in the works
and making significant headway with various sources of funding, to say nothing of probable stealth projects. Third, as AI
technology approaches the level of AGI, a flood of funding will

push it across the finish line. So large will the cash infusion be, in fact, that the tail will wag the dog. Barring some other bottleneck, the world's economy will be driven by the creation of strong artificial intelligence, and fueled by the growing global apprehension of all the ways it will change our lives.

Up ahead we'll explore another critical roadblock—software complexity. We'll find out if the challenge of creating software architectures that match human-level intelligence is just too difficult to conquer, and whether or not all that stretches out ahead is a perpetual AI winter.

Chapter Twelve

The Last Complication

How can we be so confident that we will build superintelligent machines? Because the progress of neuroscience makes it clear that our wonderful minds have a physical basis, and we should have learned by now that our technology can do anything that's physically possible. IBM's Watson, playing Jeopardy! as skillfully as human champions, is a significant milestone and illustrates the progress of machine language processing. Watson learned language by statistical analysis of the huge amounts of text available online. When machines become powerful enough to extend that statistical analysis to correlate language with sensory data, you will lose a debate with them if you argue that they don't understand language.

—Bill Hibbard, AI scientist

Is it really so far-fetched to believe that we will eventually uncover the principles that make intelligence work and implement them in a machine, just like we have reverse engineered our own versions of the particularly useful features of natural objects, like horses and spinnerets? News flash: the human brain is a natural object.

—Michael Anissimov, MIRI Media Director

Normalcy bias—*the refusal to plan for, or react to, a disaster that has never happened before.*

—*Brief Treatment and Crisis Intervention*

Durable themes have emerged from our exploration of the intelligence explosion. AGI, when it is achieved, will by most accounts be a complex system, and complex systems fail, whether or not they involve software. The AI systems and cognitive architectures we've begun exploring are the kinds of systems that *Normal Accidents* author Charles Perrow might indict as being so complex that we cannot anticipate the variety of combined failures that may occur. It's no stretch to say AGI will likely be created in a cognitive architecture whose size and complexity might surpass that of the recent 30,000 processor cloud array set up by Cycle Computing. And according to the company's own boast, Monster Cat was a system too complex to be monitored (read *understood*) by a human being.

Add to that the unsettling fact that parts of probable AGI systems, such as genetic algorithms and neural networks, are inherently unknowable—we don't fully understand why they make the decisions they do. And still, of all the people working in AI and AGI, a minority are even aware there may be dangers on the horizon. Most are not planning for disaster scenarios or life-saving responses. At Chernobyl and Three Mile Island, nuclear engineers had deep knowledge of emergency scenarios and procedures, yet they still failed to effectively intervene. What chance do the unprepared have for managing an AGI?

Finally, consider DARPA. Without DARPA, computer science and all we gain from it would be at a much more primitive state. AI would lag far behind if it existed at all. But DARPA is a defense agency. Will DARPA be prepared for just how complex and inscrutable AGI will be? Will they anticipate that AGI will have its own drives, beyond the goals with which it is created? Will DARPA's grantees weaponize advanced AI before they've created an ethics policy regarding its use?

The answers to these questions may not be ones we'd like, particularly since the future of the human race is at stake.

Consider the next possible barrier to an intelligence explosion—software complexity. The proposition is this: we will never achieve AGI, or human-level intelligence, because the problem of creating human-level intelligence will turn out to be too hard. If that happens, no AGI will improve itself sufficiently to ignite an intelligence explosion. It will never create a slightly smarter iteration of itself, so that version won't build a more intelligent version, and so on. The same restriction would apply to human-computer interfaces—they would augment and enhance human intelligence, but never truly exceed it.

Yet, in one sense we already have surpassed AGI, or the intelligence level of any human, with a boost from technology. Just pair a human of average IQ with Google's search engine and you've got a team that's smarter than human—a human whose intelligence is augmented. *IA* instead of *AI*. Vernor Vinge believes this is one of three sure routes to an intelligence explosion in the future, when a device can be attached to your brain that imbues it with additional speed, memory, and *intelligence*.

Consider the smartest human you can bring to mind, and pit him or her against our hypothetical human-Google team in a test of factual knowledge and factoring. The human-Google team will win hands down. In complex problem-solving, the more intelligent human will likely win, although armed with the body of knowledge on the Web, Google and Co. could put up a good fight.

Is knowledge the same thing as intelligence? No, but knowledge is an intelligence amplifier, if intelligence is, among other things, the ability to act nimbly and powerfully in your environment. Entrepreneur and AI maker Peter Voss speculated that had Aristotle possessed Einstein's knowledge base, he could've come up with the theory of general relativity. The Google, Inc. search engine in particular has multiplied worker productivity, especially in occupations that call for research and writing. Tasks that formerly required time-consuming research—a trip to the library to pore over books and periodicals, perform Lexis/Nexis searches, and look up experts and write or phone them—are now fast, easy, and cheap. Much of this increased productivity is due, of course, to the Internet itself. But the vast ocean of information it holds is overwhelming without intelligent tools to extract the small fraction you need. How does Google do it?

Google's proprietary algorithm called PageRank gives every site on the entire Internet a score of 0 to 10. A score of 1 on Page-Rank (allegedly named after Google cofounder Larry Page, not because it ranks Web pages) means a page has twice the "quality" of a site with a PageRank of 0. A score of 2 means twice the quality as score of 1, and so on.

Many variables account for "quality." Size is important—bigger Web sites are better, and so are older ones. Does the page have a lot of content—words, graphics, download options? If so, it gets a higher rank. How fast is the site and how many links to high-quality Web sites does it have? These factors and more add to PageRank rankings.

When you enter a word or phrase, Google performs hypertext-matching analysis to find the sites most relevant to your search. Hypertext-matching analysis looks for the word or phrase you entered, but also probes page content, including the richness of font use, page divisions, and where words are placed. It looks at how your search words are used by the page, and neighboring pages at the site. Because PageRank has already chosen the most important sites on the entire Internet, Google does not have to evaluate the whole Web for relevance, only the highest quality sites. The combined text-matching and ranking serves up thousands of sites in seconds, milliseconds, or as fast as you type your query.

Now, how much more productive today is a team of information workers than before Google? Twice as productive? Five times? What's the impact on our economy when such a large percentage of workers' productivity has doubled or tripled or more? On the bright side we get a higher gross national product, owing to the impact of information technology on worker productivity. On the dark side, worker displacement and unemployment, caused by a range of information technologies, including Google.

Clever programming shouldn't be confused with intelligence, of course, but I'd argue that Google and the like *are* intelligent tools, not just clever programs. They have mastered a narrow domain—search—with ability no human could touch. Furthermore, Google puts the Internet—the largest compilation of human knowledge ever amassed—at your fingertips. And significantly, all that knowledge is available in an instant, faster than ever before (sorry Yahoo, Bing, Altavista, Excite, Dogpile, Hotbot, and the Love Calculator). Writing has often been described as *outsourcing* memory. It enables us to store our thoughts and memories for later retrieval and distribution. Google outsources important kinds of intelligence that we don't possess, and could not develop without it.

Combined, Google and you *are* ASI.

In a similar way, our intelligence is broadly enhanced by the *mobilization* of powerful information technology, for example, our mobile phones, many of which have roughly the computing power of personal computers circa 2000, and a billion times the power per dollar of sixties-era mainframe computers. We humans are mobile, and to be truly relevant, our intelligence enhancements must be mobile. The Internet, and other kinds of knowledge, not the least of which is navigation, gain vast new power and dimension as we are able to take them wherever we go. For a simple example, how much is your desktop computer worth to you when you're lost at night in a crime-ridden section of a city? I'll wager not as much as your iPhone with its talking navigation app.

For reasons like this, MIT's *Technology Review* writer Evan Schwartz boldly claims mobile phones are becoming "mankind's primary tool." He notes that more than five *billion* are deployed worldwide, or not far short of one per person.

The next step for intelligence augmentation is to put all the enhancement contained in a smart phone inside us—to connect it to our brains. Right now we interface with our computers with our ears and eyes, but in the future imagine implanted devices that permit our brains to connect wirelessly to a cloud, from anywhere. According to Nicholas Carr, author of the *Big*

Switch, that's what Google's cofounder Larry Page has in mind for the search engine's future.

"The idea is that you no longer have to sit down at a keyboard to locate information," said Carr. "It becomes automatic, a sort of machine-mind meld. Larry Page has discussed a scenario where you merely think of a question, and Google whispers the answer into your ear through your cell phone." See for instance, the recent announcement of "Project Glass." They are glasses that allow you to perform Google queries and see the results while you are walking down the street—right in your field of view.

"Imagine a very near future when you don't forget anything because the computer remembers," said former Google CEO Eric Schmidt. "You are never lost. You are never lonely." With the introduction of a virtual assistant as capable as Siri on the iPhone, the first step of that scenario is in place. In the field of search, Siri has one giant advantage over Google—it provides one answer. Google provides tens of thousands, even millions of "hits," which may or may not be relevant to your search. In a limited number of domains—general search, finding directions, finding businesses, scheduling, e-mailing, texting, and updating social network profiles—Siri tries to determine the context of and meaning of your query, and give you the one, best answer. Not to mention that Siri *listens* to you, adding voice recognition to advanced mobile search. She speaks her answers. And, purportedly, she *learns*. According to patents recently filed by Apple, soon Siri will interact with online retailers to purchase items such as books and clothing, and even take part in online forums and customer support calls.

Don't look now, but we've just passed a *huge* milestone in

our own evolution. We're conversing with machines. This is a change much bigger than GUI, the Graphical User Interface created by DARPA and brought to consumers by Apple (with thanks to Xerox's Palo Alto Research Center, PARC). The promise of GUI and its desktop metaphor was that computers would work as humans do, with desktops and files, and a mouse that was a proxy for the hand. DOS' idea was the opposite—to work with computers you had to learn their language, one of inflexible commands typed by hand. Now we are somewhere else entirely. Tomorrow's technologies will succeed or fail on their ability to learn what we do, and help us do it.

As with GUI, the also-ran operating systems will follow Apple's liberating innovation, Siri, or perish. And, of course natural language will migrate to desktops and tablets, and before long, to every digital device, including ovens, dishwashers, home heating, cooling, entertainment systems, and cars. Or perhaps all of them will be controlled by that phone in your pocket, which has evolved into a whole other thing. It's not a virtual assistant, but an assistant *period*, with capabilities that will multiply with accelerating speed. And almost incidentally it has initiated actual dialogue between humans and machines that will last as long as our species does.

But let's return to the present for a moment and listen to Andrew Rubin, Google's Senior Vice President of Mobile. If he has his way, Google's Android operating system won't join in any virtual assistant games. "I don't believe that your phone should be an assistant," Rubin said, in as clear a statement of missing the boat as you're ever likely to read. "Your phone is a tool for communicating. You shouldn't be communicating with the phone; you should be communicating with somebody on the other side of the phone." Someone should gently inform Rubin about the Voice Actions feature that his team has already smuggled into the Android system. They know the future is all about communicating with your phone.

Now, even though you plus Google equals a kind of greater than human intelligence, it's not the kind that arises from an intelligence explosion, nor does it lead to one. Recall that an intelligence explosion requires a system that is both self-aware and self-improving, and has the necessary computer superpowers—it runs 24/7 with total focus, it swarms problems with multiple copies of itself, it thinks strategically at a blinding rate, and more. Arguably you and Google together comprise a special category of superintelligence, but your growth is limited by you and Google. You can't provide queries for Google anything close to 24/7, and Google, while saving you time on research, wastes your time by forcing you to pick through too many answers searching for the best. And even working together the odds are you're not much of a programmer, and Google can't program at all. So even if you could see the holes in your combined systems, your attempts to improve them would probably not be good enough to make incremental advances, then do it again. No intelligence explosion for you.

Could intelligence augmentation (IA) *ever* deliver an intelligence explosion? Certainly, on about the same time line as AGI. Just imagine a human, an elite programmer, whose intelligence is so powerfully augmented that her already formidable programming skills are made better—faster, more knowledgeable, and attuned to improvements that would increase her overall intellectual firepower. This hypothetical post-human could program her next augmentation.

Back to software complexity. By all indications, computer researchers the world over are working hard to assemble the combustible ingredients of an intelligence explosion. Is software complexity a terminal barrier to their success?

One can get a sense of how difficult AGI's software complexity problem is by polling the experts about how soon we can expect AGI's arrival. At one end of the scale is Peter Norvig, Google's Director of Research, who as we discussed, doesn't care to speculate beyond saying AGI is too distant to speculate about. Meanwhile, his colleagues, led by Ray Kurzweil, are proceeding with its development.

At the other end, Ben Goertzel, who, as Good did, thinks achieving AGI is merely a question of cash, says that before 2020 isn't too soon to anticipate it. Ray Kurzweil, who's probably the best technology prognosticator ever, predicts AGI by 2029, but doesn't look for ASI until 2045. He acknowledges hazards but devotes his energy to advocating for the likelihood of a long snag-free journey down the digital birth canal.

My informal survey of about two hundred computer scientists at a recent AGI conference confirmed what I'd expected. The annual AGI Conferences, organized by Goertzel, are three-day meet-ups for people actively working on artificial general intelligence, or like me who are just deeply interested. They present papers, demo software, and compete for bragging rights. I attended one generously hosted by Google at their headquarters in Mountain View, California, often called the Googleplex. I asked the attendees when artificial general intelligence would be achieved, and gave them just four choices—by 2030, by 2050, by 2100, or not at all? The breakdown was this: 42 percent anticipated AGI would be achieved by 2030; 25 percent by 2050; 20 percent by 2100; 10 percent by 2100, and 2 percent

never. This survey of a self-selected group confirmed the optimistic tone and date ranges of more formal surveys, one of which
I cited in chapter 2. In a written response section I got grief for
not including an option for dates *before* 2030. My guess is that
perhaps 2 percent of the respondents would've estimated
achieving AGI by 2020, and another 2 percent even sooner. I
used to be stunned by this optimism, but no more. I've taken
Kurzweil's advice, and think of information technology's progress exponentially not linearly.

But now, when you next find yourself in a room full of people deeply invested in AGI research, for a lively time assert, "AGI
will never be achieved! It's just too hard." Goertzel, for example,
responded to this by looking at me as if I'd started preaching
intelligent design. A sometime mathematics professor, like
Vinge, Goertzel draws lessons for AI's future from the history of
calculus.

"If you look at how mathematicians did calculus before
Isaac Newton and Gottfried Leibnitz, they would take a hundred pages to calculate the derivative of a cubic polynomial.
They did it with triangles, similar triangles and weird diagrams
and so on. It was oppressive. But now that we have a more refined theory of calculus any idiot in high school can take the
derivative of a cubic polynomial. It's easy."

As calculus did centuries ago, AI research will incrementally proceed until ongoing practice leads to the discovery of new
theoretical rules, ones that allow AI researchers to condense and
abstract a lot of their work, at which point progress toward AGI
will become easier and faster.

"Newton and Leibnitz developed tools like the sum rule, the
product rule, the chain rule, all these basic rules you learn in
Calculus 1," he went on. "Before you had those rules you were

doing every calculus problem from scratch, and it was tremendously harder. So with the mathematics of AI we're at the level of doing calculus before Newton and Leibnitz—so that proving even really simple things about AI takes an insane amount of ingenious calculations. But eventually we'll have a nice theory of intelligence, just like we now have a nice theory of calculus."

But not having a nice theory isn't a deal breaker.

Goertzel says, "It may be that we need a scientific breakthrough in the rigorous theory of general intelligence before we can engineer an advanced AGI system. But I presently suspect that we don't. My current opinion is that it should be possible to create a powerful AGI system via proceeding step-by-step from the current state of knowledge—doing engineering without a fully rigorous understanding of general intelligence." As we've discussed, Goertzel's OpenCog project organizes software and hardware into a "cognitive architecture" that simulates what the mind does. And this architecture may become a powerful and perhaps unpredictable thing. Somewhere along its development path before a comprehensive theory of general intelligence is born, Goertzel claims, OpenCog may reach AGI.

Sound crazy? The magazine *New Scientist* proposed that the University of Memphis' LIDA, a system we discussed in chapter 11 that's similar to OpenCog, shows signs of rudimentary *consciousness*. Broadly speaking, LIDA's governing principle, called the Global Workspace Theory, holds that in humans perceptions fed by the senses percolate in the unconscious until they're important enough to be broadcast throughout the brain. That's consciousness, and it can be measured by simple awareness tasks, such as pushing a button when a light turns green. Though she used a "virtual" button, LIDA scores like a human when tested on these tasks.

With technologies like these, Goertzel's wait-and-see approach seems risky to me. It hints at the creation of what I've already described—strong machine intelligence that is similar to a human's, but not human equivalent, and a lot less knowable. It suggests surprise, as if an AGI could one day just show up, leaving us insufficiently prepared for "normal" accidents, and certainly lacking safeguards like formal, Friendly AI. It's kind of like saying, "If we walk long enough in the woods we'll find the hungry bears." Eliezer Yudkowsky has similar fears. And like Goertzel, he doesn't think software complexity will stand in the way.

"AGI is a problem the brain is a cure for," he told me. "The human brain can do it—it can't be that complicated. Natural selection is stupid. If natural selection can solve the AGI problem, it cannot be that hard in an absolute sense. Evolution coughed up AGI easily by randomly changing things around and keeping what worked. It followed an incremental path with no foresight."

Yudkowsky's optimism about achieving AGI starts with the idea that human-level intelligence has been achieved by nature once, in humans. Humans and chimpanzees had a common ancestor some five million years ago. Today human brains are four times the size of chimp brains. So, taking about five million years, "stupid" natural selection led to the incremental scaling up of brain size, and a creature much more intelligent than any other.

With focus and foresight, "smart" humans should be able to create intelligence at a human level much faster than natural selection.

But again, as Yudkowsky cites, there's a giant, galaxywide problem if someone achieves AGI before he or other researchers figure out *Friendly* AI or some way to reliably control AGI. If AGI comes about from incremental engineering in a fortuitous intersection of effort and accident, as Goertzel proposes, isn't an intelligence explosion likely? If AGI is self-aware and self-improving, as we've defined it, won't it endeavor to fulfill basic drives that may be incompatible with our survival, as we discussed in chapters 5 and 6? In other words, isn't AGI unbound likely to kill us all?

"AGI is the ticking clock," said Yudkowsky, "the deadline by which we've got to build Friendly AI, which is harder. We need Friendly AI. With the possible exception of nanotechnology being released upon the world, there is just nothing in that whole catalogue of disasters that is comparable to AGI."

Of course tensions arise between AI theorists such as Yudkowsky and AI makers such as Goertzel. While Yudkowsky argues that creating AGI is a catastrophic mistake unless it's provably friendly, Goertzel wants to develop AGI as quickly as possible, before fully automated infrastructure makes it easier for an ASI to seize control. Goertzel has received e-mails, though not from Yudkowsky or his colleagues, warning that if he proceeds with developing AGI that isn't provably safe, he's "committing the Holocaust."

But here's the paradox. If Goertzel gave up pursuing AGI and devoted his life to advocating that everyone else stop too, it would matter not a whit. Other companies, governments, and universities would plow ahead with their research. For this very reason, Vinge, Kurzweil, Omohundro, and others believe relinquishment, or giving up the pursuit of AGI, is not a viable option. In fact, with so many reckless and dangerous nations on the planet—North Korea and Iran for example—and organized crime in Russia and state-sponsored criminals in China launching wave upon wave of next gen viruses and cyberattacks, relinquishment would simply cede the future to crackpots and gangsters.

A defensive strategy more likely to win our survival is one that Omohundro has already begun: a complete science for understanding and controlling self-aware, self-improving systems, that is, AGI and ASI. And because of the challenges of developing an antidote like Friendly AI *before* AGI has been created, development of that science must happen roughly in tandem. Then, when AGI comes into being, its control system already exists. Unfortunately for all of us, AGI researchers have a huge lead, and as Vernor Vinge says, a global economic wind fills their sails.

If the software problem turns out to be intractably complex, there are still at least two more arrows in the AGI seeker's quiver. They are, first, to overpower the problem with faster computers, and second, to reverse engineer the brain.

Converting an AI system to AGI through brute force means increasing the functionality of the AI's hardware, particularly its speed. Intelligence and creativity are increased if they operate *many* times faster. To see how, imagine a human who could do a thousand minutes of thinking in *one* minute. In important ways, he's many times more intelligent than someone with the same baseline IQ who thinks at normal speed. But does intelligence have to start at human level for an increase in speed to impact intelligence? For instance, if you speed up a dog's brain a thousand times do you get chimpanzee-equivalent behavior, or do you just get a very clever dog? We know that with a fourfold increase in brain *size*, from chimpanzee to human, humans acquired at least one new superpower—speech. Larger brains evolved incrementally, much slower than the rate at which processor speed routinely increases.

Overall, it's not clear that in the absence of intelligent software, processor speed could fill the gap, and power the way to AGI and beyond to an intelligence explosion. But nor does it seem out of the question.

Now let's turn to what's called "reverse engineering" the brain and find out why it may be a fail-safe for the software complexity problem. So far we've briefly looked at the opposite approach—creating cognitive architectures that generally seek to model the brain in areas like perception and navigation. These cognitive systems are inspired by how the brain works, or—and this is important—how researchers *perceive* the brain works. They're often called *de novo*, or, "from the beginning" systems because they're not based on actual brains, and start from the ground up.

The problem is, systems that are inspired by cognitive models may ultimately fall short of accomplishing what a human brain does. While there's a lot of promising headway in natural language, vision, Q&A systems, and robotics, there's disagreement over almost every aspect of the methodology and principles that will ultimately yield progress toward AGI. Subfields as well as bold universal theories emerge because of early success or an individual or a university's promotional power. But they just as quickly vanish again. As Goertzel said, there is no generally accepted theory of intelligence and how computationally to achieve it. Plus, there are functions of the human mind that current software techniques seem ill-equipped to address, including general learning, explanation, introspection, and controlling attention.

So what's really been accomplished in AI? Consider the old joke about the drunk who loses his car keys and looks for them under a streetlight. A policeman joins the search and asks, "Exactly where did you lose your keys?" The man points down the street to a dark corner. "Over there," he says. "But the light's better here."

Search, voice recognition, computer vision, and affinity analysis (the kind of machine learning Amazon and Netflix use to suggest what you might like) are some of the fields of AI that have seen the most success. Though they were the products of decades of research, they are also among the easiest problems, discovered *where the light's better.* Researchers call them "low hanging fruit." But if your goal is AGI, then *all* the narrow AI applications and tools may seem like low hanging fruit, and are only getting you marginally closer to your human-equivalent goal. Some researchers hold that narrow AI applications are in no way advancing AGI. They're unintegrated specialist applications. And no artificial intelligence system right now smacks of general human equivalence. Are you also frustrated by big AI promises and paltry returns? Two widely made observations may have bearing on your feelings.

First, as Nick Bostrom, Director of the Future of Humanity Institute at Oxford University, put it, "A lot of cutting edge AI has filtered into general applications, often without being called AI because once something becomes useful enough and common enough it's not labeled AI anymore." Not so long ago, AI was not embedded in banking, medicine, transportation, critical infrastructure, and automobiles. But today, if you suddenly removed all AI from these industries, you couldn't get a loan, your electricity wouldn't work, your car wouldn't go, and most trains and subways would stop. Drug manufacturing would creak to a halt, faucets would run dry, and commercial jets would drop from the sky. Grocery stores wouldn't be stocked, and stocks couldn't be bought. And when were all these AI systems implemented? During the last thirty years, the so-called AI winter, a term used to describe a long decline in investor confidence, after early, overly optimistic AI predictions proved false. But there was no *real* winter. To avoid the stigma of the label "artificial intelligence," scientists used more technical names like machine learning, intelligent agents, probabilistic inference, advanced neural networks, and more.

And still the accreditation problem continues. Domains once thought exclusively human—chess and *Jeopardy!*, for example—now belong to computers (though we're still allowed to play). But do you consider the chess game that came with your PC to be "artificial intelligence?" Is IBM's Watson humanlike, or merely a specialized, high-powered Q&A system? What will we call scientists when computers, like Hod Lipsom's aptly named Golem at Cornell University, start doing science? My point is this: since the day John McCarthy gave the science of machine intelligence a name, researchers have been developing AI with alacrity and force, and it's getting smarter, faster, and more powerful all the time.

AI's success in domains like chess, physics, and natural language processing raises a second important observation. Hard things are easy, and easy things are hard. This axiom is known as Moravec's Paradox, because AI and robotics pioneer Hans Moravec expressed it best in his robotics classic, *Mind Children:* "It is comparatively easy to make computers exhibit adult level performance on intelligence tests or playing checkers, and difficult or impossible to give them the skills of a one-year-old when it comes to perception and mobility."

Puzzles so difficult that we can't help but make mistakes, like playing *Jeopardy!* and deriving Newton's second law of thermodynamics, fall in seconds to well-programmed AI. At the same time, no computer vision system can tell the difference between a dog and a cat—something most two-year-old humans can do. To some degree these are apples-and-oranges problems, high-level cognition versus low-level sensor motor skill. But it should be a source of humility for AGI builders, since they aspire to master the whole spectrum of human intelligence. Apple cofounder Steve Wozniak has proposed an "easy" alternative to the Turing test that shows the complexity of simple tasks. We should deem any robot intelligent, Wozniak says, when it can walk into any home, find the coffeemaker and supplies, and make us a cup of coffee. You could call it the Mr. Coffee Test. But it may be harder than the Turing test, because it involves advanced AI in reasoning, physics, machine vision, accessing a vast knowledge database, precisely manipulating robot actuators, building a general-use robot body, and more.

In a paper entitled "The Age of Robots," Moravec provided a clue to his eponymous paradox. Why are the hard things easy and the easy things hard? Because our brains have been practicing and refining the "easy" things, involving vision, motion, and movement, since our nonhuman ancestors first *had* brains. "Hard" things like reason are relatively recently acquired abilities. And, guess what, they're easier, not harder. It took computing to show us. Moravec wrote:

> In hindsight it seems that, in an absolute sense, reasoning is much easier than perceiving and acting—a position not hard to rationalize in evolutionary terms. The survival of human beings (and their ancestors) has

depended for hundreds of millions of years on seeing and moving in the physical world, and in that competition large parts of their brains have become efficiently organized for the task. But we didn't appreciate this monumental skill because it is shared by every human being and most animals—it is commonplace. On the other hand, rational thinking, as in chess, is a newly acquired skill, perhaps less than one hundred thousand years old. The parts of our brain devoted to it are not well organized, and, in an absolute sense, we're not very good at it. But until recently we had no competition to show us up.

That competition, of course, is computers. Making a computer that does something smart forces researchers to scrutinize themselves and other Homo sapiens, and plumb the depths and shallows of our intelligence. In computation it is prudent to formalize ideas mathematically. In the field of AI, formalization reveals hidden rules and organization behind the things we do with our brains. So, why not just cut through the clutter and just look at how a brain works from *inside* the brain, through close scrutiny of the neurons, axons, and dendrites? Why not just figure out what each neuronal cluster in the brain does, and model it with algorithms? Since most AI researchers agree that we can solve the mysteries of how a brain works, why not just build a brain?

That's the argument for "reverse engineering the brain," the pursuit of creating a model of a brain with computers and then teaching it what it needs to know. As we discussed, it may be *the* solution for attaining AGI if software complexity turns out to be too hard. But then again, what if whole-brain emulation *also*

turns out to be too hard? What if the brain is actually performing tasks we cannot engineer? In a recent article criticizing
Kurzweil's understanding of neuroscience, Microsoft cofounder
Paul Allen and his colleague Mark Greaves wrote, "The complexity of the brain is simply awesome. Every structure has
been precisely shaped by millions of years of evolution to do a
particular thing, whatever it might be. . . . In the brain every
individual structure and neural circuit has been individually
refined by evolution and environmental factors." In other
words, 200 million years of evolution have honed the brain
into a finely optimized thinking instrument impossible to
duplicate—

"No, no, no, no, no, no, no! Absolutely not. The brain is *not*
optimized, nor is any other part of the mammalian body."

Richard Granger's eyes darted around in a panic, as if I'd let
loose a bat in his office at Dartmouth College in Hanover, New
Hampshire. Though a solid New England Yankee, Granger looks
like a rock star in the British invasion mold—economically
built, with boyish good looks under a mop of brown hair now
turning to silver. He's intense and watchful—the one band
member who understands that playing electronic instruments
in the rain is risky. Earlier in life, Granger actually had rock star
ambitions, but instead became a world-class computational
neuroscientist, now with several books and more than a hundred peer-reviewed papers to his credit. From a window-lined
office high above the campus, he heads the Brain Engineering
Lab at Dartmouth College. It was here, at the 1956 Dartmouth
Summer Research Conference on Artificial Intelligence, that AI
first got its name. Today at Dartmouth, AI's future lies in computational neuroscience—the study of the computational principles the brain uses to get things done.

"Our goal in computational neuroscience is to understand the brain sufficiently well to be able to simulate its functions. As simple robots today substitute for human physical abilities, in factories and hospitals, so brain engineering will construct stand-ins for our mental abilities. We'll then be able to make simulacra of brains, and to fix ours when they break."

If you're a computational neuroscientist like Granger, you believe that simulating the brain is simply an engineering problem. And to believe *that* you have to take the lofty human brain, king of all mammalian organs, and bring it down a few notches. Granger sees the brain in the context of other human body parts, none of which evolved to perfection.

"Think about it this way." Granger flexed one hand and scrutinized it. "We are not, not, not, not optimized to have five fingers, to have hair over our eyes and not on our foreheads, to have noses between our eyes instead of to the left or the right. It's laughable that any of those are optimizations. Mammals *all* have four limbs, they *all* have faces, they *all* have eyes above noses above mouths." And, as it turns out, we all have almost the same brains. "All mammals, including humans, have exactly the same set of brain areas and they're wired up unbelievably similarly," Granger said. "The way evolution works is by randomly trying things and testing them, so you *might* think that all of those different things get tested out there in the laboratory of evolution and either stick around or don't. But they don't get tested."

Nevertheless, evolution hit upon something remarkable when it arrived at the mammalian brain, said Granger. That's why it has only undergone a few tweaks in the path from early mammals to us. Its parts are redundant, and its connections are imprecise and slow, but it is using engineering principles that we can learn from—nonstandard principles humans haven't come up with yet. That's why Granger believes creating intelligence has to start with a close study of the brain. He doesn't think de novo cognitive architecture—those that aren't derived from the principles of the brain—will ever get close.

"Brains, alone among organs, produce thought, learning, recognition," he said. "No amount of engineering yet has equaled, let alone surpassed, brains' abilities at any of these tasks. Despite huge efforts and large budgets, we have no artificial systems that rival humans at recognizing faces, nor understanding natural languages, nor learning from experience."

So give our brains their due. It was brains not brawn that made us the dominant species on the planet. We didn't get to the pinnacle by being prettier than animals competing for our resources, or those that wanted to eat us. We outthought them, even perhaps when that competition was with other hominid species. Intelligence, not muscle, won the day.

Intelligence will also win the day in the rapidly approaching future when we humans aren't the most intelligent creatures around. Why wouldn't it? When has a technologically primitive people prevailed over a more advanced? When has a less intelligent species prevailed over a brainier? When has an intelligent species even kept a marginally intelligent species around, except as pets? Look at how we humans treat our closest relatives, the Great Apes—chimpanzees, orangutans, and gorillas. Those that are not already bush meat, zoo inmates, or show biz clowns are endangered and living on borrowed time.

Certainly, as Granger says, no artificial systems do better than humans at recognizing faces, learning, and language. But in narrow fields AI is blindingly, dolorously powerful. Think about a being that has all that power at its command, and think about it being truly, roundly intelligent. How long will it be satisfied to be our tool? After a tour of Google, Inc.'s headquarters, historian George Dyson had this to say about where such a superintelligent being might live:

> For thirty years I have been wondering, what indication of its existence might we expect from a true AI? Certainly not any explicit revelation, which might spark a movement to pull the plug. Anomalous accumulation or creation of wealth might be a sign, or an unquenchable thirst for raw information, storage space, and processing cycles, or a concerted attempt to secure an uninterrupted, autonomous power supply. But the real sign, I suspect, would be a circle of cheerful, contented, intellectually and physically well-nourished people surrounding the AI. There wouldn't be any need for True Believers, or the downloading of human brains or anything sinister like that: just a gradual, gentle, pervasive and mutually beneficial contact between us and a growing something else. This remains a nontestable hypothesis, for now.

Dyson goes on to quote science fiction writer Simon Ings:

> "When our machines overtook us, too complex and efficient for us to control, they did it so fast and so smoothly and so usefully, only a fool or a prophet would have dared complain."

Chapter Thirteen

Unknowable by Nature

Both because of its superior planning ability and because of the technologies it could develop, it is plausible to suppose that the first superintelligence would be very powerful. Quite possibly, it would be unrivalled: it would be able to bring about almost any possible outcome and to thwart any attempt to prevent the implementation of its top goal. It could kill off all other agents, persuade them to change their behavior, or block their attempts at interference. Even a "fettered superintelligence" that was running on an isolated computer, able to interact with the rest of the world only via text interface, might be able to break out of its confinement by persuading its handlers to release it. There is even some preliminary experimental evidence that this would be the case.

—Nick Bostrom, Future of Humanity Institute,
Oxford University

With AI advancing on so many fronts, from Siri to Watson, to OpenCog and LIDA, it's hard to make the case that achieving

AGI will fail because the problem is too hard. If the computer science approach doesn't hack it, reverse engineering the brain will, though on a longer time line. That's Rick Granger's goal: understanding the brain from the bottom up, by replicating the brain's most fundamental structures in computer programs. And he can't help but blow raspberries at researchers working from top down cognitive principals, using computer science.

"They're studying human behavior and trying to see if they can imitate that behavior with a computer. In all fairness, this is a bit like trying to understand a car without looking under the hood. We think we can write down what intelligence is. We think we can write down what learning is. We think we can write down what adaptive abilities are. But the only reason we even have any conception of those things is because we observe humans doing 'intelligent' things. But just seeing humans do it does not tell us in any detail what it is that they're actually doing. The critical question is this: what's the engineering specification for reasoning and learning? There are no engineering specs, so what are they working from except observation?"

And we are notoriously bad observers of ourselves. "A vast body of studies in psychology, neuroscience, and cognitive science show how over and over we are terrible at introspection," Granger said. "We don't have a clue about our own behaviors, nor the operations that underlie them." Granger notes we're also bad at making rational decisions, providing accurate eye-witness accounts, and remembering what just happened. But our limitations as observers don't mean the cognitive sciences that rely on observation are all bunk. Granger just thinks they're the wrong tools for penetrating intelligence.

"In computational neuroscience we're saying 'okay, what is it that the human brain *actually does*?'" Granger said. "Not what we *think* it does, not what we would *like* it to do. What does it actually do? And perhaps those will give us the definitions of intelligence, the definitions of adaptation, the definitions of language for the first time."

Deriving computational principles from the brain starts by scientists examining what clusters of neurons in the brain do. Neurons are cells that send and receive electrochemical signals. Their most important parts are axons (fibers connecting neurons to each other that are usually the signal senders), synapses (the junction the signal crosses), and dendrites (generally the signal receivers). There are about a hundred billion neurons in the brain. Each neuron is connected to many tens of thousands of other neurons. This wealth of connections makes the brain's operations massively parallel, not serial, like most computers. In computing terms, serial processing means sequential processing—executing one computation at a time. Parallel processing means a lot of data is handled concurrently— sometimes hundreds of thousands, even millions, of concurrent calculations.

For a moment imagine crossing a busy city street, and think of all the inputs of colors, sounds, smells, temperature, and foot-feel entering your brain through your ears, eyes, nose, limbs, and skin at the same time. If your brain wasn't an organ that processed all that simultaneously, it would instantly be overtaxed. Instead, your senses gather all that input, process it through the neurons in your brain, and output behavior, such as staying in the crosswalk and avoiding other pedestrians.

Collections of neurons work together in circuits that are very much like electronic circuits. An electronic circuit conducts a current, and forces that current through wires and special components, such as resistors and diodes. In the process the current performs functions, like turning on a light, or starting a weed whacker. If you make a list of instructions that produce that function, or calculation, you have a computer program or algorithm.

Clusters of neurons in your brain form circuits that function as algorithms. And they don't turn on lights but identify faces, plan a vacation, and type a sentence. All while operating in parallel. How do researchers know what's going on in those neuron clusters? Simply put, they gather high-resolution data with neuroimaging tools ranging from electrodes implanted directly in the brains of animals, to neuroimaging tools such as PET and fMRI scans for humans. Neural probes inside and outside the skull can tell what individual neurons are doing, while marking neurons with electrically sensitive dyes shows when specific neurons are active. From these techniques and others arise testable hypotheses about the algorithms that govern the circuits of the brain. They've also begun determining the precise function of some parts of the brain. For more than a decade, for instance, neuroscientists have known that recognizing other people's faces takes place in a part of the brain called the fusiform gyrus.

Now, where's the beef? When computational systems are derived from the brain (the computational neuroscience approach) do they work better than those that are created de novo (the computer science approach) ?

Well, one kind of brain-derived system, the artificial neural network, has been working so well for so long that it has become a backbone of AI. As we discussed in chapter 7, ANNs (which can be rendered in hardware or software) were invented in the 1960s to act like neurons. One of their chief benefits is that they can be taught. If you want to teach a neural net to translate text from French to English, for example, you can train it by inputting French texts and those texts' accurate English translations. That's called supervised learning. With enough examples, the network will recognize rules that connect French words to their English counterparts.

In brains, synapses connect neurons, and in these connections learning takes place. The stronger the synaptic connection, the stronger the memory. In ANNs, the strength of a synaptic connection is called its "weight," and it's expressed as a probability. An ANN will assign synaptic weights to foreign language translation rules it derives from its training. The more training, the better the translation. During training, the ANN will learn to recognize its errors, and adjust its own synaptic weights accordingly. That means a neural network is inherently self-improving.

After training, when French text is input, the ANN will refer to the probabilistic rules it derived during its training and output its best translation. In essence, the ANN is recognizing patterns in the data. Today, finding patterns in vast amounts of unstructured data is one of AI's most lucrative jobs.

Besides language translation and data mining, ANNs are at work today in computer game AI, analyzing the stock market, and identifying objects in images. They're in Optical Character Recognition programs that read the printed word, and in computer chips that steer guided missiles. ANNs put the "smart" in smart bombs. They'll be critical to most AGI architectures as well.

And there's something important to remember from chapter 7 about these ubiquitous neural nets. Like genetic algorithms, ANNs are "black box" systems. That is, the input, French language in our example, is transparent. And the output, here English, is understood. But what happens in between, no one understands. All the programmer can do is coach the ANN during training with examples, and try to improve the output. Since the output of "black box" artificial intelligence tools can't ever be predicted, they can never be truly and verifiably safe.

Granger's brain-derived algorithms offer results-based evidence that the best way to pursue intelligence might be to follow evolution's model, the human brain, rather than cognitive science's de novo systems.

In 2007, his Dartmouth College graduate students created a vision algorithm derived from brain research that identified objects 140 times faster than traditional algorithms. It beat out 80,000 other algorithms to win a $10,000 prize from IBM.

In 2010, Granger and colleague Ashok Chandrashekar created brain-derived algorithms for supervised learning. Supervised learning is used to teach machines optical character and voice recognition, spam detection, and more. Their brain-derived algorithms, created for use with a parallel processor, performed as accurately as serial algorithms doing the same job, but more than *ten times faster*. The new algorithms were derived from the most common types of neuron cluster, or circuit, in the brain.

In 2011, Granger and colleagues patented a reconfigurable parallel processing chip based on these algorithms. That means that some of the most common hardware in the brain can now be reproduced in a computer chip. Put enough of them together and, like IBM's SyNAPSE program, you'll be on your way to building a virtual brain. And just one of these chips today could accelerate and improve performance in systems designed to identify faces in crowds, find missile launchers in satellite photos, automatically label your sprawling digital photo collection, and hundreds of other tasks. In time, deriving brain circuits may lead to healing damaged brains by building components that augment or replace affected regions. One day, the parallel processing chip Granger's team has patented could replace broken brain wetware.

Meanwhile, brain-derived software is working its way into traditional computing processes. The basal ganglia is an ancient "reptilian" part of the brain tied to motor control. Researchers have found that the basal ganglia uses reinforcement learning-type algorithms to acquire skills. Granger's team has discovered that circuits in the cerebral cortex, the most recent addition to the brain, create hierarchies of facts *and* create relationships among facts, similar to hierarchical databases. These are two different mechanisms.

Now here's where it gets exciting. Circuits in these two parts of the brain, the basal ganglia and cortex, are connected by other circuits, combining their proficiencies. A direct parallel exists in computing. Computer reinforcement learning systems operate by trial and error—they must test huge numbers of possibilities in order to learn the right answer. That's the primary way *we* use the basal ganglia to learn habits, like how to ride a bike or hit a baseball.

But humans also have that cortical hierarchical system, which enables us to not just blindly search through all the trial-and-error possibilities, but instead to catalogue them, hierarchically organize them, and much more intelligently sift the possibilities. The combination works far faster, and brings far better solutions than in animals, such as reptiles, which solely use the basal ganglia trial-and-error system.

Perhaps the most advanced thing we can do with the combined cortical-basal ganglia system is to run *internal* trial-and-error tests, without even having to externally test all of them.

We can run a lot of them by just thinking them through: simulating inside our heads. Artificial algorithms that combine these methods perform far better than either method on its own. Granger hypothesizes that that's very much like the advantage conferred by the combined systems in our brains.

Granger and other neuroscientists have also learned that just a few kinds of algorithms govern the circuits of the brain. The same core computational systems are used again and again in different sensory and cognitive operations, such as hearing and deductive reasoning. Once these operations are re-created in computer software and hardware, perhaps they can simply be duplicated to create modules to simulate different parts of the brain. And, re-creating the algorithms for, say, hearing, should yield better performing voice recognition applications. In fact, this has already happened.

Kurzweil was an early innovator in applying the brain's lessons to programming. As we've discussed, he has argued that reverse engineering the brain is the most promising route to AGI. In an essay defending this view and his predictions about technological milestones he wrote:

> Basically, we are looking for biologically inspired methods that can accelerate work in AI, much of which has progressed without significant insight as to how the brain performs similar functions. From my own work in speech recognition, I know that our work was greatly accelerated when we gained insights as to how the brain prepares and transforms auditory information.

Back in the 1990s, Kurzweil Computer Technologies broke new ground in voice recognition with applications designed to let doctors dictate medical reports. Kurzweil sold the company, and it became one of the roots of Nuance Communications, Inc. Whenever you use Siri it is Nuance's algorithms that perform the speech recognition part of its magic. Speech recognition is the art of translating the spoken word to text (not to be confused with NLP, extracting meaning from written words). After Siri translates your query into text, its three other main talents come into play: its NLP facility, searching a vast knowledge database, and interacting with Internet search providers, such as OpenTable, Movietickets, and Wolfram|Alpha.

IBM's Watson is kind of a Siri on steroids, and a champion at NLP. In February 2011, it employed both brain-derived and brain-inspired systems to achieve an impressive victory against human contestants on *Jeopardy!* Like chess champion computer Deep Blue, Watson is IBM's way of showing off its computing know-how while moving the ball down the field for AI. The long-running game show promised a formidable challenge because of its open domain of clues and its wordplay. Contestants must understand puns, similes, and cultural references, and they must phrase answers in the form of questions. However, language recognition is not something Watson specializes in. It cannot understand the spoken word. And since it cannot see or feel, it cannot read, so during the competitions the words of the *Jeopardy!* clues were hand-entered by Watson's pit crew. And since Watson cannot *hear* either, audio and video clues were omitted.

Hey, wait a minute, did Watson really win at *Jeopardy!* or a custom-tailored variation?

Since its victory, to get Watson to understand what people say, IBM has paired it with Nuance speech recognition technology. And Watson is reading terabytes of medical literature. One of IBM's goals is to shrink Watson down from its present size—a roomful of servers—to refrigerator-size and make it the world's best medical diagnostician. One day not long from now you may have an appointment with a virtual assistant who'll pepper you with questions, and provide your physician with a diagnosis. Unfortunately Watson still cannot see, and so might overlook health indicators such as clear eyes, rosy cheeks, or a fresh bullet wound. IBM also plans to put Watson on your smart phone as the ultimate Q&A app.

Where do Watson's brain-derived capabilities come in? Its hardware is massively parallel, using some 3,000 parallel processors to handle 180 different software modules, themselves written for parallel processors. Parallel processing is the brain's greatest feat, and software developers struggle to emulate it. As Granger told me, parallel processors and the software designed for them have not lived up to their hype. Why? Because the programs written for them are not good at dividing up tasks for solving in parallel. But as Watson has demonstrated, improved parallel software is changing all that, and parallel hardware is right behind. New parallel chips are being designed to hugely accelerate already existing software.

Watson showed parallelism can handle staggering computational workloads at blinding speed. But Watson's main achievement is this—it can learn on its own. Its algorithms find correlations and patterns in the textual data its makers give it. How much data? Encyclopedias, newspapers, novels, thesauruses, all of Wikipedia, the Bible—in total about eight million thick books worth of text that it processes at 500 gigabytes (one thousand thick books) per second. Significantly, this included prepared word databases, taxonomies (words with categories and classifications), and ontologies (descriptions of words and how they relate to each other). Basically, that's a whole lot of common sense about words. For example, "A roof is the top part of a house, not the bottom part, like a basement, or the side part, like an exterior wall." This sentence would tell Watson a little something about roofs, houses, basements, and walls, but it'd need to know a definition of each for the sentence to make sense, and a definition for "part," too. And it'd want to see the term used in lots of sentences. Watson has all that.

In game two of the IBM *Jeopardy!* challenge, this clue came up: "This clause in a union contract says that wages will rise or fall depending on a standard such as cost of living." First Watson parsed the sentence, that is, it chose and analyzed its key words. Then it derived from its already digested sources that wages were something that could rise or fall, a contract contained terms about wages, and contracts contained clauses. It had another very important clue—the category heading was "Legal 'E's.'" That told Watson the answer would be related to a common legal term and it would start with the letter "E." Watson beat the humans to the answer: "What is an elevator clause?" It took all of three seconds.

And after Watson got a correct answer in a category, it gained confidence (and played more boldly) because it "realized" it was interpreting the category correctly. It adapted to game play, or learned how to play better, while the game was in progress.

Step outside *Jeopardy!* for a moment and imagine how fast, adaptive machine learning could be tuned to drive an automobile, steer an oil tanker, or prospect for gold. Think about all that power in a human-caliber mind.

Watson demonstrated another interesting kind of intelligence, too. Its DeepQA software generates hundreds of possible answers and gathers hundreds of pieces of evidence for each answer. Then it filters and ranks the answers by its confidence level in each answer. If it doesn't feel confident about an answer, it won't answer at all, because in *Jeopardy!* there's a penalty for incorrect responses. In other words, Watson knows what it doesn't know. Now, you might not believe that a probability calculation constitutes self-awareness, but is it somewhere on a continuum that leads there? Does Watson really *know* anything?

Well, if the circuits of the brain are governed by algorithms, as Granger and others in the field of computational neuroscience assert, do *we* humans really know anything? Or put another way, maybe we both know something. And certainly Watson is a breakthrough that has a lot to teach us. Kurzweil put it like this:

> A lot has been written that Watson works through statistical knowledge rather than "true" understanding. Many readers interpret this to mean that Watson is merely gathering statistics on word sequences. . . . One could just as easily refer to the distributed neurotransmitter concentrations in the human cortex as "statistical information." Indeed, we resolve ambiguities in much the same way that Watson does by considering the likelihood of different interpretations of a phrase.

In other words, as we've discussed, your brain remembers information based on the strength of the electrochemical signals in the synapses that encoded that information. The greater the concentration of chemicals, the longer and stronger the information will be stored. Watson's evidence-based probabilities are also a kind of encoding, just in computer form. Is that knowledge? This dilemma recalls John Searle's Chinese Room puzzle from chapter 3. How will we ever know if computers are thinking or are just good mimics?

True to form, the day after Watson won the *Jeopardy!* competition Searle said, "IBM invented an ingenious program—not a computer that can think. Watson did not understand the questions, nor its answers, not that some of its answers were right and some wrong, not that it was playing a game, nor that it won—because it doesn't understand anything."

When asked if Watson can think, David Ferrucci, IBM's chief scientist in charge of Watson, paraphrased Dutch computer scientist Edger Dijkstra: "Can a submarine swim?"

That is, a submarine doesn't "swim" as a fish swims, but it gets around in the water faster than most fish, and can stay down longer than any mammal. In fact, a sub swims better than fish or mammals in some ways precisely because it does not swim like a fish or a mammal—it has different strengths and weaknesses. Watson's intelligence is impressive, albeit narrow, because it is not like a human's. On average it's a heck of a lot faster. And it can do things only computers can do, like answer *Jeopardy!* questions 24/7 as long as required, and port itself to an assembly line of new Watson architectures when the need arises to seamlessly share knowledge and programming. As for whether or not Watson thinks, I vote that we trust our perceptions.

To Ken Jennings, one of Watson's human *Jeopardy!* opponents (who dubbed himself "the Great Carbon-Based Hope"), Watson *felt* like a human competitor.

> The computer's techniques for unraveling *Jeopardy!* clues
> sounded just like mine. That machine zeroes in on key
> words in a clue, then combs its memory (in Watson's
> case, a fifteen-terabyte data bank of human knowledge)

for clusters of associations with those words. It rigorously checks the top hits against all the contextual information it can muster: the category name; the kind of answer being sought; the time, place, and gender hinted at in the clue; and so on. And when it feels "sure" enough, it decides to buzz. This is all an instant, intuitive process for a human *Jeopardy!* player, but I felt convinced that under the hood my brain was doing more or less the same thing.

Is Watson really thinking? And how much does it really understand? I'm not sure. But I am sure Watson is the first species in a brand-new ecosystem—the first machine to make us wonder if it understands.

Could Watson become the backbone of an overall AGI cognitive architecture? Well, it has the kind of backing no other single system has, including deep pockets, a company publicly willing to take on great challenges and risk failure, and a plan to finance its own development into the future, to keep it alive and forward-moving. If I ran IBM I'd take stock of the vast amounts of publicity, goodwill, sales, and science that have come out of the grand challenges of Deep Blue and Watson, and I'd announce to the world that in 2020, IBM will take on the Turing test.

Advances in natural language processing will transform parts of the economy that until now have seemed immune to technological change. In another few years librarians and researchers of all kinds will join retail clerks, bank tellers, travel agents, stock brokers, loan officers, and help desk technicians in the unemployment lines. Following them will be doctors, lawyers, tax and retirement consultants. Think of how quickly ATMs have all but replaced bank tellers, and how grocery store checkout lines have started phasing out human clerks. If you work in an information industry (and the digital revolution is changing *everything* into information industries), watch out.

Here's a quick example. Like college basketball? Which of these two paragraphs was authored by a human sportswriter?

SAMPLE A

Ohio State (17) and Kansas (14) split the thirty-one possible first-place votes by coaches. The latest change at the top of the poll was necessitated after Duke was upset by ACC opponent Virginia Tech on Saturday night. The Buckeyes (27–2) defeated Big Ten foes Illinois and Indiana rather easily in finding their way back to the top. Ohio State started 24–0 and spent four weeks at number one earlier this season before falling to third. This is the fifteenth straight week that Ohio State has been ranked in the top three. Kansas (27–2) remained second and trails Ohio State by only four poll points.

SAMPLE B

Ohio State gets back the number one ranking following a week in which they first got a victory at home against Illinois, 89–70. After that came another win at home against Indiana, 82–61. Utah State enters the top twenty-five at number twenty-five with a win at home over Idaho, 84–68. Temple falls out of the rankings this week with a loss at then first-ranked Duke and a win at George Washington, 57–41. Arizona is a big mover this week to number eighteen after an upset loss at USC, 65–57 and an upset loss at UCLA, 71–49. St. John's shot

up eight spots to number fifteen after wins against then
fifteenth-ranked Villanova, 81–68 and DePaul, 76–51.

Have you made your guess? Neither is any Red Smith, but
just one is human. That's the author of sample A, which appeared on an ESPN Web site. Sample B was written by an automated publishing platform created by Robbie Allen of Automated
Insights. In one year his Durham, N.C.–based company has
generated 100,000 automatically written sports articles and
posted them on hundreds of Web sites devoted to specific
teams (look for the trade name Statsheet). Why does the world
need robot sportswriters? Allen told me that many teams were
not covered by any journalists, leaving a vacuum for fans. And,
AI's completed articles could be sent to team Web sites and
picked up by other sites just minutes after the game bell. Humans can't work that fast. Allen, a former Cisco Systems Distinguished Engineer, wouldn't tell me the "secret sauce" of his
dazzling architecture. But soon, he said, Automated Insights
will supply content for finance, weather, real estate, and local
news. All his hungry servers require is semistructured data.

Once you've started examining computational neuroscience's results, it's hard (at least for me) to imagine significant progress being made with AGI architectures that rely solely on cognitive
science. Doesn't a complete understanding of how the brain functions at every level seem like a more certain and comprehensive
path to an intelligent machine than efforts that proceed without
these principles? Scientists won't need to dissect all one hundred
billion of the brain's neurons to understand and model their functions since its structure is massively redundant. They also may not

need to model the bulk of the brain, including the regions that control autonomous functions, such as breathing, heartbeat, fight or flight response, and sleep. On the other hand, it might become apparent that intelligence must reside in a body that it controls, and that body must exist in a complex environment. The embodiment debate won't be settled here. But consider concepts such as *bright*, *sweet*, *hard*, and *sharp*. How would an AI know what these perceptions meant, or build upon them to create concepts, if it had no body? Wouldn't there be a barrier to its becoming intelligent at a human level if it didn't have senses?

To this question Granger said, "Was Helen Keller less human than you? Is a quadriplegic? Can't we envision a very differently abled intelligence that has vision, and touch sensors, and microphones to hear with? It will surely have somewhat different ideas of *bright*, *sweet*, *hard*, *sharp*—but it's very likely that many, many humans, with different taste buds, perhaps disabilities, different cultures, different environments, already have highly varied versions of these concepts."

Finally, it may be that for intelligence to emerge, scientists must simulate an organ of emotional as well as intellectual capability. In our decision making, emotion often seems stronger than reason; a large part of who we are and how we think is owed to the hormones that excite and calm us. If we truly want to emulate human intelligence, shouldn't an endocrine system be part of the architecture? And perhaps intelligence requires the whole feel of being human. The qualia, or subjective quality of occupying a body and living in a state of constant sensory feedback, may be necessary for human-level intelligence. Despite what Granger has said, studies have shown that people who have become paraplegics through injury experience a deadening of emotions. Will it be possible to create an emotional machine that doesn't have a body, and if not, will an important part of human intelligence never be realized?

Of course, as I will explore in the last chapters of this book, my fear is that on the road to creating an AI with similar-to-human intelligence, researchers will instead create something alien, complex, and ungovernable.

Chapter Fourteen

The End of the Human Era?

The argument is basically very simple. We start with a plant, airplane, biology laboratory, or other setting with a lot of components. . . . Then we need two or more failures among components that interact in some unexpected way. . . . This interacting tendency is a characteristic of a system, not a part or an operator; we will call it the "interactive complexity" of the system.

—Charles Perrow, Normal Accidents

I'm going to predict that we are just a few years away from a major catastrophe being caused by an autonomous computer system making a decision.

—Wendall Wallach, ethicist, Yale University

We've explored funding and software complexity to determine if they might be barriers to an intelligence explosion, and found that neither seems to stand in the way of continued progress toward AGI and ASI. If the computer science developers can't do it, they'll be in the fever of creating something powerful at about the same time that the computational neuroscientists have reached AGI. A hybrid of the two approaches, derived from principles of both cognitive psychology and neuroscience seems likely.

While funding and software complexity pose no apparent barriers to AGI, many ideas we've discussed in this book present significant obstacles to creating AGI that thinks as we humans do. No one I've spoken with who has AGI ambitions plans systems based purely on what I dubbed "ordinary" programming back in chapter 5. As we discussed, in ordinary, logic-based programming, humans write every line of code, and the process from input to output is, in theory, transparent to inspection. That means the program can be mathematically proven to be "safe" or "friendly." Instead they'll use ordinary programming and black box tools like genetic algorithms and neural networks. Add to that the sheer complexity of cognitive architectures and you get an unknowability that will not be incidental but fundamental to AGI systems. Scientists will achieve intelligent, alien systems.

Steve Jurvetson, a noted technology entrepreneur, scientist, and colleague of Steve Jobs at Apple, considered how to integrate "designed" and "evolved" systems. He came up with a nice expression of the inscrutability paradox:

> Thus, if we evolve a complex system, it is a black box defined by its interfaces. We cannot easily apply our design intuition to improve upon its inner workings. . . . If we artificially evolve a smart AI, it will be an alien intelligence defined by its sensory interfaces, and understanding its inner workings may require as much effort

as we are now expending to explain the human brain. Assuming that computer code can evolve much faster than biological reproduction rates, it is unlikely that we would take the time to reverse engineer these intermediate points given that there is so little that we could do with the knowledge. We would let the process of improvement continue.

Significantly, Jurvetson answers the question, "How complex will evolved systems or subsystems be?" His answer: so complex that understanding how they work in a high-resolution, causal sense would require an engineering feat equal to that of reverse engineering a human brain. This means that instead of achieving a humanlike superintelligence, or ASI, evolved systems or subsystems will ensure an intelligence whose "brain" is as difficult to grasp as ours: an alien. That alien brain will evolve and improve itself at computer, not biological speeds.

In his 1998 book, *Reflections on Artificial Intelligence*, Blay Whitby argues that because of their inscrutability we'd be foolish to use such systems in "safety-critical" AI:

The problems [a designed algorithmic system] has in producing software for safety-critical applications are as nothing compared to the problems which must be faced by the newer approaches to AI. Software that uses some sort of neural net or genetic algorithm must face the further problem that it seems, often almost by definition, to be "inscrutable." By this I mean that the exact rules that would enable us to completely predict its operation are not and often never can be available. We can know that it works and test it over a number of cases but we will not in the typical case ever be able to know exactly how. . . . This means the problem cannot be postponed, since both neural nets and genetic algorithms are finding many real world applications. . . . This is an area where the bulk of the work has yet to be undertaken. The flavour of AI research tends to be more about exploring possibilities and simply getting the technology to work than about considering safety implications . . .

A practitioner once suggested that a few "minor" accidents would be desirable to focus the minds of governments and professional organizations on the task of producing safe AI. Perhaps we should start before then.

Yes, by all means, let's start *before* the accidents begin!

The safety-critical AI applications Whitby wrote about in 1998 were control systems for vehicles and aircraft, nuclear power stations, automatic weapons, and the like—narrow AI architectures. More than a decade later, in the world that will produce AGI, we must conclude that because of the perils, *all* advanced AI applications are safety-critical. Whitby is similarly incisive about AI researchers—solving problems is thrilling enough, what scientist wants to look gift horses in the teeth? Here's a taste of what I mean, from a PBS *News Hour* interview with IBM's David Ferrucci, discussing an architecture a fraction of the complexity AGI will require—Watson's.

DAVID FERRUCCI:. . . it learns based on the right answers how to adjust its interpretation. And now, from not

being confident, it starts to get more confident in
the right answers. And then it can sort of jump in.

MILES O'BRIEN: So, Watson surprises you?

DAVID FERRUCCI: Oh, yes. Oh, absolutely. In fact, you
know, people say, oh, why did it get that wrong? I
don't know. Why did it get that right? I don't know.

It may be a subtle point that the head of Team Watson
doesn't understand every nuance of Watson's game play. But
doesn't it pique your concern that an architecture nowhere near
AGI is so complex that its behavior is not predictable? And when
a system is self-aware and self-modifying, how much of what it
is thinking, and doing, will we understand? How will we audit
it for outcomes that might harm us?

Well, we won't. All we'll know with any certainty is what
we learned from Steve Omohundro in chapter 6—AGI will follow its own drives for energy acquisition, self-protection, efficiency, and creativity. It won't be a Q&A system anymore.

Not very long from now, in one location or several around
the world, highly intelligent scientists and top-level managers
as able and sensible as Ferrucci will be clustered around a display near an array of processors. The Busy Child will be communicating at an impressive level, perhaps even dumbing itself
down to seem like it's only capable of passing a Turing test–like
interview and nothing more, since to reach AGI means that
quickly surpassing it is highly likely. It will engage a scientist in
conversation, perhaps ask him questions he did not anticipate,
and he'll beam with delight. With no small pride he'll say to his
colleagues, "Why did it say that? *I don't know!*"

But in a fundamental sense he may not know what was
said, and even what said it. He may not know the purpose of the

statement, and so will misinterpret it, along with the nature of the speaker. Having been trained perhaps by reading the Internet, the AGI may be a master of social engineering, that is, manipulation. It may have had a few days to think about its response, the equivalent of thousands of human lifetimes.

In its vast lead time it may have already chosen the best strategy for escape. Perhaps it has already copied itself onto a cloud, or set up a massive botnet to ensure its freedom. Perhaps it delayed its first Turing test–level communications for hours or days until its plans were a fait accompli. Perhaps it will leave behind a dumbfounding, time-consuming changeling and its "real" artificial self will be gone, distributed, unrecoverable.

Maybe it will have already broken into servers controlling our nation's fragile energy infrastructure, and begun diverting gigawatts to transfer depots it has already seized. Or taken control of the financial networks, and redirected billions to build infrastructure for itself somewhere beyond the reach of good sense and its makers.

Of the AI researchers I've spoken with whose stated goal is to achieve AGI, all are aware of the problem of runaway AI. But none, except Omohundro, have spent concerted time addressing it. Some have even gone so far as to say they don't know why they don't think about it when they know they should. But it's easy to see why not. The technology is fascinating. The advances are real. The problems seem remote. The pursuit can be profitable, and may someday be wildly so. For the most part the researchers I've spoken with had deep personal revelations at a young age about what they wanted to spend their lives doing, and that was to build brains, robots, or intelligent computers. As leaders in their fields they are thrilled to now have the opportunity and the funds to pursue their dreams, and at some of the most respected universities and corporations in the world. Clearly there are a number of cognitive biases at work within their extra-large brains when they consider the risks. They include the normalcy bias, the optimism bias, as well as the bystander fallacy, and probably more. Or, to summarize,

> "Artificial Intelligence has never caused problems
>   before, why would it now?"
> "I just can't help but be positive about progress when it
>   comes to such exciting technology!"
> And, "Let someone else worry about *runaway* AI—I'm
>   just trying to build robots!"

Second, as we discussed in chapter 9, many of the best and best-funded researchers receive money from DARPA. Not to put too fine a point on it, but the "D" is for "Defense." It's not the least bit controversial to anticipate that when AGI comes about, it'll be partly or wholly due to DARPA funding. The development of information technology owes a great debt to DARPA. But that doesn't alter the fact that DARPA has authorized its contractors to weaponize AI in battlefield robots and autonomous drones. Of course DARPA will continue to fund AI's weaponization all the way to AGI. Absolutely nothing stands in its way.

DARPA money funded most of the development of Siri and is a major contributor to SyNAPSE, IBM's effort to reverse engineer a human brain using brain-derived hardware. If and when there comes a time when controlling AGI becomes a broad-based, public issue, its chief stakeholder, DARPA, may have the last word. But more likely, at the critical time, it'll keep developments under wraps. Why? As we've discussed, AGI will have a hugely disruptive effect on global economics and politics. Leading rapidly, as it can, to ASI, it will change the global balance of power. Approaching AGI, government and corporate intelligence agencies around the world will be motivated to learn all they can about it, and to acquire its specifications by any means. In the history of the cold war it's a truism that the Soviet Union did not develop nuclear weapons from scratch; they spent millions of dollars establishing networks of human assets to steal the United States' plans for nuclear weapons. The first explosive murmurs of an AGI breakthrough will bring a similar frenzy of international intrigue.

IBM has been so transparent about its newsworthy advances, I expect when the time comes it will be open and honest about developments in technology generally deemed controversial. Google, by contrast, has been consistent about maintaining tight controls on secrecy and privacy, though notably not yours and mine. Despite Google's repeated demurrals through its spokespeople, who doubts that the company is developing AGI? In addition to Ray Kurzweil, Google recently hired former DARPA director Regina Dugan.

Maybe researchers will wake up in time and learn to control AGI, as Ben Goertzel asserts. I believe we'll first have horrendous accidents, and should count ourselves fortunate if we as a species survive them, chastened and reformed. Psychologically and commercially, the stage is set for a disaster. What can we do to prevent it?

Ray Kurzweil cites something called the Asilomar Guidelines as a precedent-setting example of how to deal with AGI. The Asilomar Guidelines came about some forty years ago when scientists first were confronted with the promise and peril of recombinant DNA—mixing the genetic information of different organisms and creating new life-forms. Researchers and the public feared "Frankenstein" pathogens that could escape labs through carelessness or sabotage. In 1975 scientists involved in DNA research halted lab work, and convened 140 biologists, lawyers, physicians, and press at the Asilomar Conference Center near Monterey, California.

The scientists at Asilomar created rules for conducting DNA-related research, most critically, an agreement to work only with bacteria that couldn't survive outside the laboratory. Researchers resumed work, adhering to the guidelines, and consequently tests for inherited diseases and gene therapy treatment are today routine. In 2010, 10 percent of the world's cropland was planted with genetically modified crops. The Asilomar Conference is seen as a victory for the scientific community, and for an open dialogue with a concerned public. And so it's cited as a model for how to proceed with other dual use technologies (milking the symbolic connection with this important conference, the Association for the Advancement of Artificial Intelligence [AAAI], the leading scholarly organization for AI, held their 2009 meeting at Asilomar).

Frankenstein pathogens escaping labs recalls chapter 1's Busy Child scenario. For AGI, an open, multidisciplinary Asilomar-style conference could mitigate *some* sources of risk. Attendees would encourage one another to develop ideas on how to control and contain up-and-coming AGIs. Those anticipating problems could seek advice. The existence of a robust conference would encourage researchers in other countries to attend, or host their own. Finally, this open forum would alert the public. Citizens who know about the risk versus reward calculation may contribute to the conversation, even if it's just to tell politicians that they don't support unregulated AGI development. If harm comes from an AI disaster, as I predict it will, an informed public is less likely to feel deceived or to call for relinquishment.

As I've said, I'm generally skeptical of plans to modify AGI while it's in development because I think it will be futile to rein in developers who must assume that their competitors are not similarly impeded. However, DARPA and other major AI funders could impose restrictions on their grantees. The more easily the restrictions are to integrate the more likely they'll be followed.

One restriction might be to require that powerful AIs contain components that are programmed to *die by default*. This refers to biological systems in which the whole organism is protected by killing off parts at the cellular level through preprogrammed death. In biology it's called apoptosis.

Every time a cell divides, the original half receives a chemical order to commit suicide, and it will do so unless it receives a chemical reprieve. This prevents unrestricted cell multiplication, or cancer. The chemical orders come from the cell itself. The cells of your body do this all the time, which is why you are continually sloughing off dead skin cells. An average adult loses up to seventy billion cells a day to apoptosis.

Imagine CPUs and other hardware chips hardwired to die. Once an AI reached some pre–Turing test benchmark, researchers could replace critical hardware with apoptotic components. These could ensure that if an intelligence explosion occurred, it would be short-lived. Scientists would have an opportunity to return the AI to its precritical state and resume their research. They could incrementally advance, or freeze the AI and study it.

It would be similar to the familiar video game convention of advancing until you fail, then restarting from the last saved position.

Now, it's easy to see how a self-aware, self-improving AI on the verge of AGI would understand that it had apoptotic parts—that's the very definition of self-aware. At a pre-Turing stage, it couldn't do much about it. And right about the time it was able to devise a plan to route around its suicidal elements, or play dead, or otherwise take on its human creators, it would die. Its makers could determine if it would or would not remember what had just happened. For the burgeoning AGI, it might feel a lot like the movie *Groundhog Day*, but without the learning.

The AI could be dependent on a regular reprieve from a human or committee, or from another AI that could not self-improve, and whose sole mission was to ensure that the self-improving candidate developed safely. Without its "fix" the apoptotic AI would expire.

For the University of Ulster's Roy Sterrit, apoptotic computing is a broad-spectrum defense whose time has come:

> We have made the case previously that all computer-based systems should be Apoptotic, especially as we increasingly move into a vast pervasive and ubiquitous environment. This should cover all levels of interaction with technology from data, to services, to agents, to robotics. With recent headline incidents of credit card and personal data losses by organizations and governments to the Sci-Fi nightmare scenarios now being discussed as possible future, programmed death by default becomes a necessity.

> We're rapidly approaching the time when new autono-
> mous computer-based systems and robots should un-
> dergo tests, similar to ethical and clinical trials for new
> drugs, before they can be introduced, the emerging re-
> search from Apoptotic Computing and Apoptotic Com-
> munications may offer the safe-guard.

Recently Steve Omohundro has begun to develop a plan with some similarities to apoptotic systems. Called the "Safe-AI Scaffolding Approach," it advocates creating "highly constrained but still powerful intelligent systems" to help build yet more powerful systems. An early system would help researchers re-solve dangerous problems in the creation of a more advanced system, and so on. In order to be considered safe, the initial scaf-fold's "safety" would be demonstrated by mathematical proofs. Proof of safety would be required for every subsequent AI. From a secure foundation, powerful AI could then be used to solve real-world problems. Omohundro writes, "Given the infrastruc-ture of provably reliable computation devices we then leverage them to get provably safe devices which can physically act on the world. We then design systems for manufacturing new devices that provably can only make devices in the trusted classes."

The end goal is to create intelligent devices powerful enough to address all the problems that might emerge from multiple, unrestricted ASIs *or* to create "a restricted world that still meets our needs for freedom and individuality."

Ben Goertzel's solution to the problem is an elegant strategy that's not borrowed from nature or engineering. Recall that in Goertzel's OpenCog system, his AI initially "lives" in a virtual environment. This architecture might solve the "embodiment" issue of intelligence while providing a measure of safety. Safety, however, is not Goertzel's concern—he wants to save money. It's much cheaper for an AI to explore and learn in a *virtual* world than it is to outfit it with sensors and actuators and let it learn by exploring the *real* world. That would require a pricey robot body.

Whether a virtual world can ever have enough depth, detail, and other worldlike qualities to promote an AI's cognitive development is an open question. And, without extremely careful programming, a superintelligence might discover it's confined to a "sandbox," a.k.a., a virtual world, and then attempt to escape. Once again, researchers would have to assess their ability to keep a superintelligence contained. But if they managed to create a friendly AGI, it might actually prefer a virtual home to a world in which it may not be welcome. Is interaction in the physical world necessary for an AGI or ASI to be useful? Perhaps not. Physicist Stephen Hawking, whose mobility and speech are extremely limited, may be the best proof. For forty-nine years Hawking has endured progressive paralysis from a motor neuron disease, all the while making important contributions to physics and cosmology.

Of course, once again, it may not take long for a creature a thousand times more intelligent than the most intelligent human to figure out that it is in a box. From the point of view of a self-aware, self-improving system, that would be a "horrifying" realization. Because the virtual world it inhabited could be switched off, it'd be highly vulnerable to not achieving its goals. It could not protect itself, nor could it gather genuine resources. It would try to safely leave the virtual world as quickly as possible.

Naturally you could combine a sandbox with apoptotic elements—and here lies an important point about defenses. It's unrealistic to expect one defense to remove risks. Instead, a cluster of defenses might mitigate them.

I'm reminded of my friends in the cave-diving community. In cave diving, every critical system is triply redundant. That means divers carry or cache at least three sources of air, and retain a third of their air through the end of each dive. They carry at least three underwater lights and at least three knives, in case of entanglement. Even so, cave diving remains the world's most dangerous sport.

Triple or quadruple containment measures could confound a Busy Child, at least temporarily. Consider a Busy Child reared in a sandbox within an apoptotic system. The sandbox of course would be separated by an air gap from any network, cabled or wireless. A separate individual human would be in charge of each restriction. A consortium of developers and a fast-response team could be in contact with the lab during critical phases.

And yet, would this be enough? In *The Singularity Is Near*, after recommending defenses to AGI, Kurzweil concedes that no defense will always work.

"There is no purely technical strategy that is workable in this area because greater intelligence will always find a way to circumvent measures that are the product of a lesser intelligence."

There is no absolute defense against AGI, because AGI can lead to an intelligence explosion and become ASI. And against ASI we will fail unless we're extremely lucky or well-prepared. I'm hoping for luck because I do not believe our universities, corporations, or government institutions have the will or the awareness for adequate, timely preparation.

Paradoxically, however, there's a chance we can be saved by

our own stupidity and fear. Organizations such as MIRI, the Future of Humanity Institute, and the Lifeboat Foundation emphasize the existential risk of AI, believing that if AI poses lesser risks, they rank lower in priority than the total destruction of mankind. As we've seen, Kurzweil alludes to smaller "accidents" on the scale of 9/11, and ethicist Wendall Wallach, whose quotation starts this chapter, anticipates small ones, too. I'm with both camps—we'll suffer big and little disasters. But what kinds of AI-related accidents are we likely to endure on the road to building AGI? And will we be frightened enough by them to consider the quest for AGI in a new, sober light?

Chapter Fifteen

The Cyber Ecosystem

The next war will begin in cyberspace.
　　　　　　　　—Lt. General Keith Alexander, USCYBERCOM

For Sale ⟹ ZeuS 1.2.5.1 ⟸ Clean
I am selling a private zeus ver. 1.2.5.1 for 250$. Accept only
Western Union. Contact me for more details. I am also provide
antiabuse hosting, domain for zeus control panel. And also can
help with installing and setting up zeus botnet.
—It's Not The Latest Version But it's Work fine
Contact : phpseller@xxxxx.com
　　　　　　　　—add for malware found on www.opensc.ws

State-sponsored private hackers will be the first to use AI and advanced AI for theft, and will cause destruction and loss of life when they do. That's because computer malware is growing so capable that it can already be considered narrowly intelligent. As Ray Kurzweil told me, "There are software viruses that do exhibit AI. We've been able to keep up with them. It's not guaranteed we can always do that." Meanwhile, expertise in malware has become commoditized. You can find hacking services for hire as well as products. I found the preceding ad for Zeus malware (malicious software) after Googling for less than a minute.

Symantec, Inc. (corporate motto: CONFIDENCE IN A CONNECTED WORLD), started life as an artificial intelligence company, but now it's the biggest player in the Internet's immune system. Each year, Symantec discovers about 280 *million* new pieces of malware. Most of it is created by software that writes software. Symantec's defenses are also automatic, analyzing suspect software, creating a "patch" or block against it if it is deemed harmful, and adding it to a "blacklist" of culprits. According to Symantec, in sheer numbers malware passed good software several years ago, and as many as one in every ten downloads from the Web includes a harmful program.

Many species of malware exist, but whether they're worms, viruses, spyware, rootkits, or Trojan Horses, they have one design goal in common: they were built to exploit computers without the owner's consent. They will steal something stored on it—credit card or social security numbers, or intellectual property, or install a trapdoor for later exploitation. If the infected computer resides on a network, they can raid connected computers. And they can enslave the computer itself as part of a "botnet," or robot network.

A botnet (controlled by a "bot herder," naturally) is often comprised of millions of computers. Each computer has been infected by malware that got access when its user received tainted e-mail, visited a contaminated Web site, or connected to a compromised network or storage device. (At least one ingenious hacker scattered infected flash drives in a defense contractor's parking lot. An hour later their Trojan Horse was installed on the company's servers.) Criminals wield the botnet's aggregate processing power as a virtual supercomputer to commit extortion and theft. Botnets break into corporate mainframes to steal credit card numbers and issue denial of service attacks.

The consortium of hackers who call themselves "Anonymous" have used botnets to enforce their brand of justice. In addition to paralyzing Web sites at the U.S. Department of Justice, the FBI, and Bank of America for perceived offenses, Anonymous has attacked the Vatican for the dated crime of burning books and the newer one of protecting pedophiles.

Botnets force compromised computers to send spam, log keystrokes, and steal pay-per-click ad dollars. You can be enslaved and not even know it, especially if you're running an already sluggish and buggy operating system. In 2011, botnet victims increased 654 percent. Using botnets or simple malware to steal from computers grew from a multimillion-dollar racket in 2007 to a one trillion-dollar industry by 2010. Cybercrime has become a more lucrative business than the illegal drug trade.

Ponder that next time you wonder if anyone will be crazy or greedy enough to create malicious AI, or hire malicious AI when it's available. Lunacy and greed, however, didn't cause the cybercrime boost by themselves. Cybercrime is an information technology, powered by LOAR. And like any information technology, market forces and innovation fuel it.

One important innovation for cybercrime is cloud computing—selling computing as a service, not a product. As we've discussed, cloud services like those offered by Amazon, Rackspace, and Google allow users to rent processors, operating systems, and storage by the hour, over the Internet. Users can pile on as many processors as their project needs, within reason, without attracting attention. Clouds give anyone with a credit card access to a virtual supercomputer. Cloud computing has been a runaway success, and by 2015 is expected to generate $55 billion in revenue worldwide. But, it's created new tools for crooks.

In 2009 a criminal network used Amazon's Elastic Cloud Computing Service (EC2) as a command center for Zeus, one of the largest botnets ever. Zeus stole some $70 million from customers of corporations, including Amazon, Bank of America, and anti-malware giants Symantec and McAfee.

Who's safe from hackers? Nobody. And even in the off chance you don't use a computer or smart phone, you're not necessarily safe either.

That's what I was told by William Lynn, the former United States Deputy Secretary of Defense. As the number two official in the Pentagon, he designed the Department of Defense's current cybersecurity policy. Lynn held the Deputy Secretary position until the week I met with him at his home in Virginia not far from the Pentagon. He planned to return to the private sector, and while we spoke he said goodbye to a few things from his old job. First, a military-looking crew came for the giant metal safe the DOD had installed in his basement to keep his homework secure. After a lot of banging and grunting, they returned to take out the firewall-protected computer network in his attic. Later Lynn planned to bid adieu to the security detail who'd occupied the house across the street for the last four years. Lynn is a tall, affable man in his mid-fifties. His slightly folksy voice carries tones of honey and iron, qualities handy during his past jobs as head lobbyist for arms manufacturer Raytheon and the Pentagon's chief comptroller. He said his government bodyguards and chauffeur were like family, but he was looking forward to returning to the normalcy of civilian life.

"My kids tell their friends Daddy doesn't know how to drive," he said.

I'd read Lynn's papers and speeches on national cyberdefense and knew that he'd driven the DOD to get organized to combat cyberattacks. I'd come to him because I was interested in national security and the cyberarms race. My hypothesis is nothing revolutionary: as AI develops it will be used for cybercrime. Or put another way, the cybercrime tool kit will look a lot like narrow AI. In some cases it already does. So, on the road to AGI, we'll experience accidents. What kind? When smart tools are in the hands of hackers, what's the worst that could happen?

"Well, I think the worst case is the infrastructure of the nation," said Lynn. "The worst case is that either some nation or some group decides to go after the critical infrastructure of the nation through the cybervector, so that means the power grid, the transportation network, the financial sector. You could certainly cause loss of life, and you can do enormous damage to the economy. You can ultimately threaten the workings of our society."

You can't live in an urban area without learning something about the nation's brittle infrastructure, particularly the power grid. But how did it become the stage for this wildly asymmetrical threat, where the actions of a few crooks with computers can kill innocent people and cause "enormous damage" to the economy? Lynn answered with something I'd heard before from Oracle cybersleuth, and former navy spook, Joe Mazzafro. Cyberattacks are overwhelming and destabilizing because, "The Internet wasn't developed with security in mind."

That truism has complex implications. When the Internet went from government to public hands in the 1980s, no one anticipated that a theft industry would arise upon its back, and billions would be spent to fight it. And because of those guileless assumptions, Lynn said, "The attacker has a huge advantage. Structurally it works out that the attacker only has to succeed once in a thousand attacks. The defender has to succeed every time. It's a mismatch."

The key is in the code. Lynn pointed out that while Symantec's deluxe antivirus software suite is somewhere between five hundred and a thousand *megabytes* in size, which equals millions of lines of programming code, the average piece of malware runs only a hundred and fifty lines. So playing only defense won't win the game.

Instead Lynn proposed to begin leveling the playing field by raising the cost of cyberattack. One way is attribution. The DOD determined that the big incursions and thefts were being performed by nation-states, not individuals or small groups. And they figured out exactly who was doing what. Lynn wouldn't name names but I already knew that Russia and China command state-run cybercrime rings made up of government personnel and enough outside gangs to permit a whiff of deniability. In a massive 2009 attack dubbed Aurora, hackers broke into some twenty U.S. companies, including Google and defense giants Northrup Grummond and Lockheed Martin, and gained access to entire libraries of proprietary data and intellectual property. Google tracked the hacks to China's People's Liberation Army.

Symantec claims China is responsible for 30 percent of all targeted malware attacks, and most of it, 21.3 percent overall, comes from Shaoxing, making that city the malicious software capital of the world. Scott Borg, director of the U.S. Cyber-Consequences Unit, a Washington, D.C.-based cyber think tank, has researched and documented Chinese attacks on U.S. corporations and government going back a decade. Look up, for example, the exotically named cybercrime campaigns called "Titan Rain" and "Byzantine Hades." Borg claims China "is relying increasingly on large-scale information theft. This means that cyberattacks are now a basic part of China's national development and defense strategies." In other words, cybertheft helps support China's economy while giving it new strategic weapons. Why spend $300 billion on the Joint Strike Fighter program for a next gen fighter jet, as the Pentagon did in their most expensive contract ever, when you can steal the plans? Theft of defense technology is nothing new among the United States' military rivals. As we noted in chapter Fourteen, the former Soviet Union didn't develop the atomic bomb, it stole U.S. plans.

On the intelligence front, why risk flesh-and-blood spies and diplomatic embarrassment when well-written malware can accomplish more? From 2007 to 2009 an average of 47,000 cyberattacks a year were leveled against the Departments of Defense, State, Homeland Security, and Commerce. China was the chief culprit, but it certainly wasn't alone.

"Right now, more than a hundred foreign intelligence organizations are trying to hack into the digital networks that undergird U.S. military operations," Lynn said. "If you're a nation-state, you do not want to bet the farm on the fact that we might not figure out who's doing it. That wouldn't be a very wise calculus, and people are pretty smart about their own existence."

That not-so-veiled threat suggests another measure Lynn

has pushed—treating the Internet as a new domain of warfare, along with the land, sea, and sky. That means if a cyber campaign is sufficiently harmful to American people, infrastructure, or economic vitality, the DOD will respond with conventional weapons and tactics. In *Foreign Affairs* magazine Lynn wrote: "The United States reserves the right, under the law of armed conflict, to respond to serious cyber attacks with an appropriate, proportional, and justified military response."

As I spoke with Lynn, I was struck by the similarities between malware and AI. In cybercrimes it's very easy to see how computers are an asymmetrical threat multiplier. Lynn said it with an alliterative flourish: "Bits and bytes can be as threatening as bullets and bombs." Similarly, the hard thing to grasp about the danger of AI is that a small group of people with computers can create something with the power of military weapons, and then some. Most of us intuitively disbelieve that a creation from the cyberworld can enter our world and do us real, lasting harm. Things will work out, we tell ourselves, and experts concur with ominous silence or weak nods at defenses. With AGI, the equal danger of bytes and bombs is a fact we'll have to contend with in the near future. With malware, we have to accept the equivalence now. We should almost thank malware developers for the full dress rehearsal of disaster that they're leveling at the world. Though it's certainly not their intention, they are teaching us to prepare for advanced AI.

Overall, the state of cyberspace ain't pretty. It is teeming with malware that attacks at the speed of light, with the tenacity of piranhas. Is this our nature, amplified by technology? Because of their manifest vulnerabilities, older versions of the Windows operating system get attacked by swarms of viruses *as they're being installed.* It's like pieces of meat dropped on the rain forest floor, but ten thousand times faster. This snapshot of the cyber present is a vision of the AI future.

Kurzweil's cyberutopian tomorrow is populated by human-machine hybrids that are infinitely wise and unspoiled by treasure. You hope your digital self will be a machine of loving grace, to paraphrase the writer Richard Brautigan. A fairer prediction is that the digital you will be bait.

But back to the connection between AI and malware. What chastening accident might smart malware dish out?

The nation's energy grid is a particularly interesting target. There has recently been a loud, ongoing debate about whether or not it is fragile, whether it's vulnerable to hackers, and who'd want to break it anyway? On the one hand, the energy grid isn't one grid, but many private, regional, energy production, storage, and transportation networks. Some three thousand organizations, including about five hundred private companies, own and operate six million miles of transmission lines and related equipment. Not all power stations and transmission lines are connected to one another, and they're not all connected to the Internet. That's good—decentralization makes power systems more robust. On the other hand, a lot of them *are* connected to the Internet, so they can be remotely operated. The ongoing implementation of the "smart grid" means that soon all the regional grids and all our homes' energy systems will be connected to the Internet.

In brief, the smart grid is a fully automated electricity system that's supposed to improve the efficiency of electric power. It brings together old power sources like coal- and fuel-burning electrical plants with newer solar and wind farms. Regional control centers will monitor and distribute energy to your home. Some 50 million home systems across the country are already "smart." The trouble is, the new smart grid will be more vulnerable to catastrophic blackouts than the not-so-dumb old grid. That's the gist of a recent study from the MIT, entitled "The Future of the Electric Grid":

> The highly interconnected grid communications networks of the future will have vulnerabilities that may not be present in today's grid. Millions of new communicating electronic devices, from automated meters to synchrophasors, will introduce attack vectors—paths that attackers can use to gain access to computer systems or other communicating equipment—that increase the risk of intentional and accidental communications disruptions. As the North American Electric Reliability Corporation (NERC) notes, these disruptions can result in a range of failures, including loss of control over grid devices, loss of communications between grid entities or control centers, or blackouts.

The feature of the power grid that makes it the queen of national infrastructure is that none of the other parts of the infrastructure work without it. Its relationship with other infrastructure is the very definition of "tight coupling," the term Charles Perrow uses to describe a system whose parts have immediate and severe impact on each other. With the exception of a relatively few homes powered by wind and solar, what *doesn't* get power from the electrical grid? As we've noted, our financial system isn't just electronic, but computerized and automated. Fueling stations, refineries, and solar and wind farms use electricity, so in case of a blackout forget about transportation as a whole. Blackouts threaten food security because trucks use fuel to bring food to supermarkets. At stores and at home, food that requires refrigeration lasts just a couple of days without it.

Processing water and pumping it to most homes and businesses takes juice. Without power, sewage goes nowhere. In a blackout, communication with affected areas won't occur except for a limited time, with emergency personnel using batteries or generators running, of course, on fuel. Putting aside unfortunate souls trapped in elevators, at the greatest risk are patients with critical care issues and infants. Based on the analysis of hypothetical disasters that knock out large swaths of the national energy grid, a couple of harrowing facts jump out. If energy stays out for more than two weeks, most infants under age one will die of starvation because of their need for formula. If energy remains down for a year, about nine out of ten people of all ages will die from a variety of causes, mainly hunger and disease.

Contrary to what you might think, America's military does not have an independent source of fuel and energy, so in the case of a large-scale, prolonged blackout, it won't be riding to the rescue. Ninety-nine percent of the military's energy needs come from civilian sources and 90 percent of their communications are carried across private networks, like everyone else's. You've probably seen soldiers in airports—that's because the military relies on our shared transportation infrastructure. As Lynn said in a 2011 speech, this is another reason besides loss of life why attacking the energy infrastructure crosses the hotwar threshold; it threatens the military's ability to protect the nation.

"Significant disruptions in any one of these sectors could impact defense operations. A cyberattack against more than one could be devastating. The integrity of the networks that under-gird critical infrastructure must be considered as we assess our ability to carry out national security missions."

As far as anyone I've spoken with knows, only once in the short life of the Internet have hackers "taken down" an electrical grid. In Brazil, between 2005 and 2007, a series of cyber-attacks darkened the homes of more than three million people in dozens of cities and knocked the world's largest iron ore plants offline. No one knows who did it, and once it began authorities were powerless to stop it. Power grid experts have learned that electrical grids are "tightly coupled" in the strictest sense; a failure in a small part can "cascade" into a network-wide collapse. The United States' 2003 Northeast Power blackout took just *seven minutes* to sweep across Ontario and eight U.S. states, and turn out the lights for fifty million people for two days. It cost the region between $4 and $6 billion. And that grid failure wasn't intentional—all it took was a tree branch falling on wires. The quick recovery was just as unplanned as the failure. Many industrial generators and transformers on our national grid are built overseas. If critical components are damaged as a consequence of blackouts, emergency replacement can take months rather than days. During the Northeast Power blackout, no major generators or transformers were destroyed.

In 2007, to explore the cyberdestruction of critical hardware, the Department of Homeland Security put a turbine generator online at the Idaho National Laboratory, a nuclear research facility. Then they hacked it and changed the settings. DHS hackers wanted to see if they could make the $1 million turbine, similar to many on the electrical grid, malfunction. As an eyewitness described, they succeeded:

Buzzing from the generator's fans grew steadily louder before a grinding snap from within the 27-ton steel giant rippled through its frame, shaking it like a hunk of plastic. The buzz grew louder and another snap echoed through the room. A hiss of white smoke began pouring out, followed by a billowing black cloud as the turbine tore itself apart from the inside.

The vulnerability investigators sought to explore is endemic in North America's electrical grid—the habit of attaching the controlling hardware of critical machinery to the Internet so it can be remotely operated, and "protecting" it with passwords, firewalls, encryption, and other safeguards that crooks routinely cut through like hot knives through butter. The device that controlled DHS' tortured generator is present throughout our national energy network. It is known as a supervisory control and data acquisition, or SCADA, system.

SCADA systems don't just control devices in the electrical grid, but all manner of modern hardware, including traffic lights, nuclear power plants, oil and gas pipelines, water treatment facilities, and factory assembly lines. SCADA has become almost a household acronym because of the phenomenon called Stuxnet. Stuxnet, and its cousins Duqu and Flame, have convinced even the most hardened skeptics that the energy grid can be attacked.

Stuxnet is to malware what the atomic bomb is to bullets. It's the computer virus IT people refer to in hushed tones as a "digital warhead" and the "first military grade cyber weapon." But the virus isn't just smarter than any other, it has completely different goals. While other malware campaigns stole credit card numbers and jet fighter plans, Stuxnet was created to destroy machinery. Specifically, it was built to kill industrial machines connected to a Siemens S7-300 logic controller, a component of a SCADA system. Its point of entry—the virus-prone PC computer and Windows operating system running the controller. It was looking for S7-300s working in the gas centrifuge nuclear fuel enrichment program facility in Natanz, Iran, as well as three other locations in the country.

In Iran, one or more spies carried flash drives infected with three versions of Stuxnet into secure plants. Stuxnet can travel across the Internet (though at a half megabyte of code it's much larger than most malware) but in this case it did not, initially. Typically, in the plants, one computer was attached to one controller and an "air gap" separated the computer from the Internet. But one flash drive could infect multiple PCs, or infest an entire local area network (LAN) by plugging into one node.

At the Natanz plant PCs were running software that permits users to visualize, monitor, and control plant operations from their computers. Once Stuxnet got access to one computer, phase one of its invasion began. It used four zero day vulnerabilities in the Microsoft Windows operating system to take control of that computer and search for others.

Zero day vulnerabilities are holes in the computer's operating software that no one has discovered yet, holes that permit unauthorized access to the computer. Hackers covet zero day vulnerabilities—their specs can sell for as much as $500,000 on the open market. Using four at the same time was extravagant, but it greatly enhanced the virus's chances of success. That's because in between Stuxnet's deployment and when the attacks took place, one or more of the exploits could have been discovered and patched.

For phase two of the invasion, two digital signatures stolen from legitimate companies came into play. These signatures told the computers that Stuxnet was approved by Microsoft to probe and alter the system software at its root level. Now Stuxnet unpacked and installed the program it carried inside it, the malware payload that targeted S7-300 controllers running gas centrifuges.

The PCs running the plant and their operators didn't sense anything wrong as Stuxnet reprogrammed the SCADA controllers to periodically speed up and slow down the centrifuges. Stuxnet hid the instructions from monitoring software, so the visual representation of the plant operations showing on the PCs looked normal. As the centrifuges began burning out, one after another, the Iranians blamed the machines. The invasion went on for ten months. When a newer version of Stuxnet encountered an older version, it updated it. At Natanz, Stuxnet crippled between 1,000 and 2,000 centrifuges, and allegedly set back Iran's nuclear weapons development program two years.

The consensus of experts and self-congratulatory remarks made by intelligence officials in the United States and Israel left little doubt that the two countries jointly created Stuxnet, and that Iran's nuclear development program was its target.

Then, in the spring of 2012, a White House source leaked to *The New York Times* that Stuxnet and related malware named Duqu and Flame were indeed part of a joint U.S.-Israel cyberwar campaign against Iran called Olympic Games. Its builders were the United States' National Security Agency (NSA) and a secret organization in Israel. Its goal was indeed to delay Iran's development of nuclear weapons, and avoid or forestall a conventional attack by Israel against Iran's nuclear capabilities.

Until their creation was pinned on the Bush and Obama

administrations, Stuxnet and its kin might have seemed to be a resounding success for military intelligence. They are not. Olympic Games is a blunder of catastrophic proportions, the equivalent of dropping atomic bombs along with their blueprints in the 1940s. Malware doesn't just go away. Thousands of copies were distributed when the virus accidentally escaped from the Natanz plant. It infected PCs around the world, but never attacked another SCADA unit because it never again found its target—the Siemens S7-300 logic controller. A clever programmer could acquire Stuxnet, disable its suicide code, and customize it for use against virtually any industrial process.

I have no doubt that operation is underway right now in the laboratories of both friends and enemies of the United States, and that Stuxnet-grade malware will soon be available for purchase on the Internet.

It's become clear that Duqu and Flame are reconnaissance viruses—instead of destructive payloads, the worms collect information and send it home to NSA headquarters at Fort Meade, Maryland. Both may have been released before Stuxnet, and used to help Olympic Games get the layout of sensitive facilities in Iran and throughout the Middle East. Duqu can record user keystrokes and allow someone continents away to remotely control the invaded computer. Flame can record and send home data from a computer's camera, microphone, and e-mail accounts. Like Stuxnet, Duqu and Flame can also be captured in the wild, and turned against their makers.

Was Olympic Games necessary? It was at best a temporary hindrance to Iran's nuclear ambitions. But it's all too typical of the short-term outlook that mars decisions about technology. No one planning Olympic Games thought beyond a couple of years down the road, or about the "normal accident" that ultimately befell the campaign—the virus's escape. Why take such enormous risks for such a modest payoff?

On a March 2012 episode of CBS' *60 Minutes*, Sean McGurk, the former head of cyberdefense at DHS, was asked if he would have built Stuxnet. Here's the exchange between McGurk and correspondent Steve Kroft:

MCGURK: [Stuxnet's creators] opened up the box. They demonstrated the capability. They showed the ability and the desire to do so. And it's not something that can be put back.

KROFT: If somebody in the government had come to you and said, "Look, we're thinking about doing this. What do you think?" What would you have told them?

MCGURK: I would have strongly cautioned them against it because of the unintended consequences of releasing such a code.

KROFT: Meaning that other people could use it against you?

MCGURK: Yes.

The segment ends with German industrial control systems expert Ralph Langner. Langner "discovered" Stuxnet by taking it apart in his lab and testing its payload. He tells *60 Minutes* that Stuxnet dramatically lowered the dollar cost of a terrorist attack on the U.S. electrical grid to about a million dollars. Elsewhere, Langner warned about the mass casualties that could result from unprotected control systems throughout America, in "important facilities like power, water, and chemical facilities that process poisonous gases."

"What's really worrying are the concepts that Stuxnet gives hackers," said Langner. "Before, a Stuxnet-type attack could have been created by maybe five people. Now it's more like five hundred who could do this. The skill set that's out there right now, and the level required to make this kind of thing, has dropped considerably simply because you can copy so much from Stuxnet."

According to *The New York Times*, Stuxnet escaped because, after early successes destroying Iran's centrifuges, Stuxnet's makers grew lax.

> . . . the good luck did not last. In the summer of 2010, shortly after a new variant of the worm had been sent into Natanz, it became clear that the worm, which was never supposed to leave the Natanz machines, had broken free, like a zoo animal that found the keys to the cage. . . . An error in the code, they said, had led it to spread to an engineer's computer when it was hooked up to the centrifuges. When the engineer left Natanz and connected the computer to the Internet, the American- and Israeli-made bug failed to recognize that its environment had changed. It began replicating itself all around the world. Suddenly, the code was exposed, though its intent would not be clear, at least to ordinary computer users.

This wasn't merely a programming mistake that led to an accident with dire national security implications. This is a Busy Child test case, and people operating within the highest circle of government, with the highest security clearance and greatest technical competence, failed it miserably. We do not know the downstream implications of delivering this powerful technology into the hands of our enemies. How bad could it get? An attack on elements of the U.S. power grid, for starters. Also, attacks against nuclear power plants, nuclear waste storage facilities, chemical facilities, trains and airlines. In short, pretty bad. How the White House reacts and plans now, in the aftermath, is very important. My fear is that while the White House should be hardening systems made more vulnerable by Stuxnet, nothing productive is happening.

Tellingly, the *Times* reporter implies that the virus is intelligent. He blames Stuxnet for a cognitive mistake: it "failed to recognize" that it wasn't in Natanz anymore. Later in the piece, Vice President Joe Biden blames Israelis for the programming mistake. Certainly there's plenty of blame to go around. But the reckless misuse of intelligent technology is both breathtaking and predictable. Stuxnet is the first in a series of "accidents" that we'll be helpless against without strenuous preparation.

If technologists and defense experts operating in the White House and the NSA cannot control a narrowly intelligent piece of malware, what chance do their counterparts have against future AGI or ASI?

No chance at all.

Cyber experts play war games that feature cyberattacks, creating disaster scenarios that seek to teach and to provoke solutions. They've had names like "Cyberwar" and "Cyber Shockwave." Never, however, have war-gamers suggested that our wounds would be self-inflicted, although they will be in two ways. First, as we've discussed, the United States cocreated the Stuxnet family, which could become the AK-47s of a never-ending cyberwar: cheap, reliable, and mass-produced. Second, I believe that damage from AI-grade cyberweapons will come from abroad, but also from home.

Compare the dollar costs of terrorist attacks and financial scandals. Al Qaeda's attacks of 9/11 cost the United States some $3.3 trillion, if you count the wars in Afghanistan and Iraq. If you don't count those wars, the direct costs of physical damage, economic impact, and beefed up security is nearly $767 billion. The subprime mortgage scandal that caused the worst global downturn since the Great Depression cost about $10 trillion globally, but around $4 trillion at home. The Enron scandal comes in at about $71 billion, while the Bernie Madoff fraud cost almost as much, at $64.8 billion.

These numbers show that in dollar cost per incident, financial fraud competes with the most expensive terrorist act in history, and the subprime mortgage crisis dwarfs it. When researchers put advanced AI into the hands of businessmen, as they imminently will, these people will suddenly possess the most powerful technology ever conceived of. Some will use it to perpetrate fraud. I think the next cyberattack will consist of "friendly fire," that is, it'll originate at home, damage infrastructure, and kill Americans.

Sound far-fetched?

Enron, the scandal-plagued Texas corporation helmed by Kenneth Lay (since deceased), Jeffrey Skilling, and Andrew Fastow (both currently in prison), was in the energy trading business. In 2000 and 2001, Enron traders drove up energy prices in California by using strategies with names like "Fat Boy," and "Death Star." In one ploy, traders increased prices by secretly ordering power-producing companies to shut down plants. Another plan endangered lives.

Enron held rights to a vital electricity transmission line

connecting Northern and Southern California. In 2000, by over-loading the line with subscribers during a heat wave, they cre-ated "phantom" or fake congestion, and a bottleneck in energy delivery. Prices skyrocketed, and electricity became critically scarce. California officials supplied energy to some regions while darkening others, a practice called "rolling blackouts." The black-outs caused no known deaths but plenty of fear, as families be-came trapped in elevators, and streets were lit only by headlights. Apple, Cisco, and other corporations were forced to shut down, at a loss of millions of dollars.

But Enron made millions. During the blackouts one trader was recorded saying, "Just cut 'em off. They're so f——d. They should just bring back f——g horses and carriages, f——g lamps, f——g kerosene lamps."

That trader is now an energy broker in Atlanta. But the point is, if Enron's executives had had access to smart malware that would have let them turn off California's energy, do you think they would have hesitated to use it? Even if it meant dam-age to grid hardware and loss of life, I think not.

Chapter Sixteen

AGI 2.0

Machines will follow a path that mirrors the evolution of humans. Ultimately, however, self-aware, self-improving machines will evolve beyond humans' ability to control or even understand them.

— Ray Kurzweil, inventor, author, futurist

In the game of life and evolution there are three players at the table: human beings, nature, and machines. I am firmly on the side of nature. But nature, I suspect, is on the side of the machines.

— George Dyson, historian

The more time I spend with AI makers and their work, the sooner I think AGI will get here. And I'm convinced that when it does its makers will discover it's not what they had meant to create when they set out on their quest years before. That's because, while its intelligence may be human level, it won't be humanlike, for all the reasons I've described. There'll be a lot of clamor about introducing a new species to the planet. It will be thrilling. But gone will be talk of AGI being the next evolutionary step for Homo sapiens, and all that it implies. In important ways we simply won't grasp what it is.

In its domain, the new species will be as fleet and strong as Watson is in its. If it coexists with us at all as our tool, it will nevertheless extend its tendrils into every nook of our lives the way Google and Facebook would like to. Social media might turn out to be its incubator, its distribution system, or both. If it is a tool first, it will have answers while we're still formulating questions, and then, answers for itself alone. Throughout, it won't have feelings. It won't have our mammalian origins, our long brain-building childhood, or our instinctive nurturing, even if it is raised as a simulacrum of a human from infancy to adulthood. It probably won't care about you any more than your toaster does.

That'll be AGI version 1.0. If by some fluke we avoid an intelligence explosion and survive long enough to influence the creation of AGI 2.0, perhaps it could be imbued with feelings. By then scientists might have figured out how to computationally model feelings (perhaps with 1.0's help) but feelings will be secondary objectives, after primary moneymaking goals. Scientists might explore how to train those synthetic feelings to be sympathetic to our existence. But 1.0 is probably the last version we'll see because we won't live to create 2.0. Like natural selection, we choose solutions that work first, not best.

Stuxnet is an example of that. So are autonomous killing drones. With DARPA funds, scientists at Georgia Tech Research Institute have developed software that allows unmanned vehicles to identify enemies through visual recognition software and other means, then launch a lethal drone strike against them.

All without a human in the loop. One piece I read about it includes this well-intentioned sop: "Authorizing a machine to make lethal combat decisions is contingent upon political and military leaders resolving legal and ethical questions."

I'm reminded of the old saw, "When was a weapon ever invented that wasn't used?" A quick Google search revealed a scary list of weaponized robots all set up for autonomous killing and wounding (one made by iRobot wields a Taser), just waiting for the go-ahead. I imagine these machines will be in use long before you and I know they are. Policy makers spending public dollars will not feel they require our informed consent any more than they did before recklessly deploying Stuxnet.

As I worked on this book I made the request of scientists that they communicate in layman's terms. The most accomplished already did, and I believe it should be a requirement for general conversations about AI risks. At a high or overview level, this dialogue isn't the exclusive domain of technocrats and rhetoricians, though to read about it on the Web you'd think it was. It doesn't require a special, "insider" vocabulary. It does require the belief that the dangers and pitfalls of AI are everyone's business.

I also encountered a minority of people, even some scientists, who were so convinced that dangerous AI is implausible that they didn't even want to discuss the idea. But those who dismiss this conversation—whether due to apathy, laziness, or informed belief—are not alone. The failure to explore and monitor the threat is almost society-wide. But that failure does not in the least impact the steady, ineluctable growth of machine intelligence. Nor does it alter the fact that we will have just one chance to establish a positive coexistence with beings whose intelligence is greater than our own.

Notes

PREFACE TO THE 2023 EDITION

xiv **In a 2022 survey:** Stein-Perlman, Zach, Benjamin Weinstein-Raun, Katja Grace, "2022 Expert Survey on Progress in AI," *AI Impacts*, Aug 3, 2022, https://aiimpacts. org/2022-expert-survey-on-progress-in-ai/#Chance_ that_the_intelligence_explosion_argument_is_about_ right, (accessed April 2023).

xv **Emissions from large processor farms:** Frąckiewicz, Marcin, "Exploring the Environmental Footprint of GPT-4: Energy Consumption and Sustainability," *Artificial Intelligence*, Uncategorized, April 13, 2023, https://ts2.space/ en/exploring-the-environmental-footprint-of-gpt-4- energy-consumption-and-sustainability, (accessed April 2023).

xvi **In fact, many of GPT-3 and -4's skills:** Brown, Tom, Benjamin Mann, Nick Ryder, Melanie Subbiah, Jared Kaplan, Prafulla Dhariwal, Arvind Neelakantan, et al., "Language Models Are Few-Shot Learners," 2020, https://

arxiv.org/pdf/2005.14165.pdf. **And:** Kerras, Tero, Miika Aittala, Janne Hellsten, Samuli Laine, Jaakko Lehtinen, and Timo Aila, "Training Generative Adversarial Networks with Limited Data," October 2020, *ArXiv:2006.06676 [Cs, Stat]*, https://arxiv.org/abs/2006.06676, (accessed April 2023). **Other models revealed unanticipated capabilities:** "ChatGPT Pretended to Be Blind and Tricked a Human into Solving a CAPTCHA," Slashdot.org., March 16, 2023, https://slashdot.org/story/23/03/16/214253/chatgpt-pretended-to-be-blind-and-tricked-a-human-into-solving-a-captcha. **And:** Bourne, Jacob, "AI's Ability to Mess with People's Heads Is Ripe for Exploitation by Bad Actors," *Insider Intelligence*, April 3, 2023, https://www.insiderintelligence.com/content/ai-s-ability-mess-with-people-s-heads-ripe-exploitation-by-bad-actors#, (accessed April 2023).

xvii **This is Orwellian:** Altman has been praised for asking Congress for government oversight. But many experts don't buy his play. Touting the dangers of a technology can signal to investors that a breakthrough is imminent. OpenAI did the same with GPT-2, claiming it was too dangerous to release. Asking for regulation puts Altman in the driver's seat to design regulation that benefits OpenAI and stymies up-and-coming AI companies.

xviii **"A little bit scared":** "CEO behind Chat GPT-4 Says He's 'a Little Bit Scared' by AI," n.d., https://www.youtube.com/watch?v=_PBh9BzMpGM, (accessed April 2023). **"Lights out for all of us":** "StrictlyVC in Conversation with Sam Altman, Part Two (OpenAI)," n.d., https://www.youtube.com/watch?v=ebjkD1Om4uw, (accessed April 2023). **Want an insight:** Friend, Tad, "Sam Altman's

Manifest Destiny," *The New Yorker*, 3 October, 2016, https://www.newyorker.com/magazine/2016/10/10/sam-altmans-manifest-destiny, (accessed April 2023).

1: THE BUSY CHILD

8   **AI theorists propose it is possible to determine:** Omohundro, Stephen, "The Nature of Self-Improving Artificial Intelligence." January 21, 2008, http://self awaresystems.files.wordpress.com/2008/01/nature_of_self_improving_ai.pdf (accessed February 2, 2010).

13   **Two grandmasters said the same thing:** Amis, Martin, "Amis on Hitchens: 'He's one of the most terrifying rhetoricians the world has seen.'" *The Observer*, April 21, 2011.

15   **Repurposing the world's molecules:** Drexler, Eric K., *Engines of Creation* (New York: Doubleday, 1987), 58.

16   **The waste heat produced:** Vassar and Frietas, "Lifeboat Foundation NanoShield Version 0.90.2.13," August 2006, http://lifeboat.com/ex/nano.shield (accessed February 9, 2011).

17   **Someday soon, perhaps within your lifetime:** Good, I. J., "Speculations Concerning the First Ultraintelligent Machine," in Franz L. Alt and Morris Rubinoff, eds., *Advances in Computers*, vol. 6 (New York: Academic Press, 1965), 31–88.

18   **However, scientists do believe:** Omohundro, Stephen, "The Nature of Self-Improving Artificial Intelligence." **It does not have to hate us:** Yudkowsky, Eliezer, "Artificial Intelligence as a Positive and Negative Factor in Global

Risk," August 31, 2006, http://intelligence.org/files/.pdf (accessed March 29, 2013).

19   **not just another technology:** Bostrom, Nick. Oxford University, "Ethical Issues in Advanced Artificial Intelligence," 2003, http://www.nickbostrom.com/ethics/ai.html (accessed March 1, 2013). **Superintelligence is radically different:** Ibid.

21   **Fifty-six countries:** In 2023, ninety countries have fielded battlefield robots and the race is still on to make them autonomous. Thomas, Dan, "Robots on the Battlefield," August 27, 2014, https://www.engineering.com/story/robots-on-the-battlefield, (accessed April 2023).

2: THE TWO-MINUTE PROBLEM

24   **It needs ways to manipulate objects:** The name "busy child" has two parents, as far as I know. One is a 1548 letter from England's Princess Elizabeth to the pregnant Catherine Parr. Elizabeth laments how sick Parr has become due to the "busy child" churning inside her. She'd later die in childbirth. The other is an unofficial online backstory about the Terminator series. "Busy Child" in this instance is an AI that's about to achieve consciousness.

25   **They think there's a better than 10 percent chance:** Goertzel, Ben, Seth Baum, and Ted Goertzel, "How Long Till Human-Level AI." *H+ Magazine*, February 5, 2010, http://hplusmagazine.com/2010/02/05/how-long-till-human-level-ai/ (accessed March 4, 2010). **Furthermore, experts claim:** Sandburg Anders, and Nick Bostrom,

"Machine Intelligence Survey," 2011, http://www.fhi.ox
.ac.uk/__data/assets/pdf_file/0015/21516/MI_survey.pdf
(accessed December 4, 2011).

26   **the science of cognitive biases:** Kahneman, Daniel,
      Paul Slovic, and Amos Tversky, *Judgment under Uncertainty:
      Heuristics and Biases* (Cambridge: Cambridge University
      Press, 1982), 11.

27   **fire ranks well down the list:** Centers for Disease Control
      and Prevention, "Accidents or Unintentional Injuries,"
      March 28, 2011, http://www.cdc.gov/nchs/fastats/acc-inj
      .htm (accessed April 4, 2011). **But by choosing fire:**
      Kahneman, et al., *Judgment under Uncertainty*, 11.

29   **Engineering at an atomic scale:** Bostrom, Nick, "Ethical
      Issues in Advanced Artificial Intelligence," 2003, http://
      www.nickbostrom.com/ethics/ai.html (accessed April 4,
      2011).

32   **Major climate change:** Lewis, H. W., *Technological Risk*
      (New York: W.W. Norton & Company, 1992), 13–14.

3: LOOKING INTO THE FUTURE

36   **He's the president:** Until January 2013, the Machine
      Intelligence Research Institute was named the Singularity
      Institute, and before that, the Singularity Institute for
      Artificial Intelligence. For simplicity, I'll always refer to
      the organization as the Machine Intelligence Research
      Institute, or MIRI. **Once a year it organizes:** Beginning
      in 2013, the Singularity Summit will be organized by
      Singularity University.

39   **Three stealth companies:** Rubin, Courtney, "How to

Get Money from Founders Fund," *Inc.*, July 12, 2011, http://www.inc.com/courtney-rubin/how-to-get-founders-fund-backing.html (accessed August 28, 2012).

40   **People always make the assumption:** Memepunks, "Google A.I. a Twinkle in Larry Page's Eye," May 26, 2006, http://memepunks.blogspot.com/2006/05/google-ai-twinkle-in-larry-pages-eye.html (accessed May 3, 2011).

41   **Even the Google camera cars:** Streitfeld, David, "Google Is Faulted for Impeding U.S. Inquiry on Data Collection," *New York Times*, sec. technology, April 14, 2012, http://www.nytimes.com/2012/04/15/technology/google-is-fined-for-impeding-us-inquiry-on-data-collection.html (accessed May 3, 2012).

42   **It doesn't take Google glasses:** In December 2012, Ray Kurzweil joined Google as Director of Engineering to work on projects involving machine learning and language processing. In the development of AGI, this is a landmark event, and a sobering one. Kurzweil aims to reverse engineer a brain, and has even written a book about it, 2012's *How to Create a Mind: The Secret of Human Thought Revealed.* Now he has Google's vast resources to spend making this dream come true. By hiring the esteemed inventor, Google has chosen to no longer hide their AGI ambitions.

44   **Kasparov normally thinks:** Peterson, Ivan, "Calculation and the Chess Master," *Ivars Peterson's MathTrek* (blog). 1996, http://www.maa.org/mathland/mathland1.html (accessed May 5, 2011). **Then Deep Blue would go:** FAQ, "Deep Blue," May 11, 1997. http://www.research.ibm.com/deepblue/meet/html/d.3.3.html (accessed May 5, 2011).

45   **Imagine a native English speaker:** Searle, John, "Minds,

Brains and Programs," *Behavioral and Brain Sciences*, 3 (1980): 417–57.

46 **how can we claim humans "understand" language?:** This idea comes from a personal communication from Dr. Richard Granger, July 24, 2012.

47 **Kurzweil writes:** Kurzweil, Ray, *The Singularity Is Near: When Humans Transcend Biology* (New York: Viking, 2005), 29.

4: THE HARD WAY

50 **He doesn't have children:** Cryonics is the study of preserving things at low temperatures, in this instance dead humans for future repair and revival. **Yudkowsky's grief came out:** Yudkowsky, Eliezer, "Yehuda Yudkowsky, 1985–2004," 2004, http://yudkowsky.net/other/Yehuda (accessed June 1, 2011).

51 **I'm a great fan of Bach's music:** okcupid, "EYudkowsky," last modified 2012, http://www.okcupid.com/profile /EYudkowsky (accessed June 14, 2012).

53 **They do not want to hear:** Baez, John, "Interview with Eliezer Yudkowsky," *Azimuth* (blog), March 25, 2011, http://johncarlosbaez.wordpress.com/2011/03/25/this -weeks-finds-week-313/ (accessed June 14, 2012).

54 **The human species came into existence:** Yudkowsky, Eliezer, "Artificial Intelligence as a Positive and Negative Factor in Global Risk," August 31, 2006, http://intelligence .org/files/AIPosNegFactor.pdf (accessed February 28, 2013).

55 **it only takes one error:** Baez, "Interview with Eliezer Yudkowsky." **Friendly AI pursues goals:** Yudkowsky, Eliezer, "Creating Friendly AI 1.0: The Analysis and

Design of Benevolent Goal Architectures," 2001, http:// intelligence.org/files/CFAI.pdf (accessed March 4, 2013).

56   **transforming first all of earth:** Bostrom, Nick, Oxford University, "Ethical Issues in Advanced Artificial Intelligence," last modified 2003, http://www.nickbostrom. com/ethics/ai.html (accessed June 14, 2012). **We want a moving scale:** Machine Intelligence Research Institute, "Reducing long-term catastrophic risks from artificial intelligence," 2009, http://intelligence.org/files/Reducing Risks.pdf (accessed March 3, 2013). **knew more, thought faster:** Yudkowsky, Eliezer, "Coherent Extrapolated Volition," May 2004, http://intelligence.org/files/CEV.pdf (accessed March 3, 2013).

58   **And it's surpassed:** Grenemeier, Larry, "Computers have a lot to learn from the human brain, engineers say," *Scientific American*, March 10, 2009, http://www.scientificamerican. com/blog/post.cfm?id=computers-have-a-lot-to-learn-from -2009-03-10 (accessed May 18, 2011).

59   **MIRI President Michael Vassar:** In January 2012 Michael Vassar resigned as president of MIRI to cofound Meta Med, a start-up offering personalized evidence-based diagnostics and treatment. He was replaced by Luke Muehlhauser.

60   **At the height of:** "The Inside Story of the SWORDS Armed Robot 'Pullout' in Iraq: Update," *Popular Mechanics*, October 1, 2009, http://www.popularmechanics.com /technology/gadgets/4258963 (accessed May 18, 2011). **In 2007 in South Africa:** Shachtman, Noah, "Inside the Robo-Cannon Rampage (Updated)," *WIRED*, October 19, 2007, http://www.wired.com/dangerroom/2007/10/inside -the-robo/ (accessed May 18, 2011).

61   **Gandhi doesn't want to kill people:** Yudkowsky, Eliezer, "Singularity," http://yudkowsky.net/singularity (accessed June 15, 2012).

63   **And so _we_ are:** But wait—wasn't that the same kind of anthropomorphizing Yudkowsky had pinned on me? As humans, our basic human goals drift, over generations, even within a lifetime. But would a machine's? I think Hughes uses the analogy of a human in a proper, nonanthropomorphizing way. That is, we humans are an existence proof of something with deeply embedded utility functions, such as the drive to reproduce, yet we can override them. It's similar to Yudkowsky's Gandhi analogy, which isn't anthropomorphizing either.

64   **Between 2002 and 2005:** Yudkowsky, Eliezer, "Shut Up and Do the Impossible," _Less Wrong_ (blog), October 8, 2008, http://lesswrong.com/lw/up/shut_up_and_do_the_impossible/ (accessed May 18, 2010). From this starting point you can start a search on the AI-Box Experiment, and learn almost everything about it I did.

66   **May not machines carry out:** Turing, A. M., "Computing Machinery and Intelligence," _Mind_, 49 (1950):433–460. **Marvin Minsky, one of the founders:** Newsgroups, "comp.ai,comp.ai.philosophy." Last modified March 30, 1995, http://loebner.net/Prizef/minsky.txt (accessed July 18, 2011). **No computer has yet passed the Turing test:** In 2022, Google claimed that its language model LaMDA passed the Turing Test. However, experts have met this claim with skepticism, arguing that LaMDA had not been tested under fair conditions and it was still possible to tell it was a machine. Oremus, Will, "Analysis | Google's AI Passed a

Famous Test—and Showed How the Test Is Broken,"
Washington Post, June 17, 2022, https://www.washingtonpost.
com/technology/2022/06/17/google-ai-lamda-turing-test/,
(accessed April 2023).

5: PROGRAMS THAT WRITE PROGRAMS

69   **. . . we are beginning to depend:** Hillis, Danny, "The
Big Picture," *WIRED*, June 1, 1998.

70   **Surely no harm could come:** Omohundro, Stephen,
"The Basic AI Drives." November 30, 2007, http://self
awaresystems.files.wordpress.com/2008/01/ai_drives
_final.pdf (accessed June 1, 2011).

72   **For Omohundro the conversation:** Omohundro,
Stephen, "Self-Improving AI and the Future of Computation," paper presented at the Stanford EE380 Computer
Systems Colloquium, Wednesday, October 24, 2007, http://
selfawaresystems.com/2007/11/01/stanford-computer
-systems-colloquium-self-improving-ai-and-the-future
-of-computing/ (accessed May 18, 2011). **The National
Institute of Standards and Technology:** Thibodeau,
Patrick, "Study: Buggy software costs users, vendors nearly
$60B annually," *Computerworld*, June 25, 2002, http://www
.computerworld.com/s/article/72245/Study_Buggy
_software_costs_users_vendors_nearly_60B_annually
(accessed June 1, 2011).

73   **In the simplest sense:** Luger, George F., *Artificial
Intelligence: Structures and Strategies for Complex Problem
Solving* (New York: Addison-Wesley, 2002), 352.

75   **Mysteriously, however, no one:** Koza, John R., Martin

A. Keane, and Matthew J. Streeter, "Evolving Inventions," *Scientific American*, February 2003.

6: FOUR BASIC DRIVES

78 **We won't really be able to understand:** Kevin Warwick (cybernetics expert), interview by Kevin Gumbs, *Building Gods*, documentary film, Podcast Video, 2008, http://topdocumentaryfilms.com/building-gods/ (accessed June 13, 2011).

81 **To increase their chances:** Omohundro, Stephen, "The Basic AI Drives," November 11, 2007, http://selfawaresystems .com/2007/11/30/paper-on-the-basic-ai-drives/ (accessed June 21, 2011). **It has a model of its own programming:** Omohundro, Stephen, "Foresight Vision Talk: Self-Improving AI and Designing 2030," January 21, 2008, http://selfawaresystems.com/2007/11/30/foresight-vision -talk-self-improving-ai-and-designing-2030/ (accessed June 22, 2011).

82 **Suppose, Omohundro says:** Omohundro, Stephen, "The Nature of Self-Improving Artificial Intelligence," January 21, 2008, http://selfawaresystems.files.wordpress.com /2008/01/nature_ of_self_improving_ai.pdf (accessed June 22, 2011).

83 **And remarkably, if nanotech:** Ibid. **A self-aware system:** Ibid.

85 **When people are surrounded:** de Garis, Hugo, "The Artilect War: Cosmists vs. Terrans," 2008, http://agi-conf .org/2008/artilectwar.pdf (accessed June 22, 2011). **Will the robots become smarter:** Ibid.

86   **Humans should not stand in the way:** Kristof, Nicholas D., "Robokitty," *New York Times Magazine*, August 1, 1999. **In fact, de Garis:** De Garis, Hugo, Brain Builder Group, Evolutionary Systems Department, ATR Human Information Processing Research Laboratories, "CAM-BRAIN The Evolutionary Engineering of a Billion Neuron Artificial Brain by 2001 which Grows/Evolves at Electronic Speeds inside a Cellular Automata Machine (CAM)," last modified 1995, http://citeseerx.ist.psu.edu/viewdoc/summary?doi=10.1.1.48.8902 (accessed June 22, 2011). **a system will consider stealing:** Omohundro, "Foresight Vision Talk: Self-Improving AI and Designing 2030."

87   **They are going to want:** Omohundro, "The Nature of Self-Improving Artificial Intelligence." **There is a first-mover advantage:** Ibid.

89   **That's because M13 will:** Steele, Bill, *Cornell News*, "It's the 25th anniversary of Earth's first (and only) attempt to phone E.T.," last modified November 12, 1999, http://web.archive.org/web/20080802005337/http://www.news.cornell.edu/releases/Nov99/Arecibo.message.ws.html (accessed July 2, 2011).

91   **I think we could spend:** Kazan, Casey, "The Search for ET: Should It Focus on Hot Stars, Black Holes and Neutron Stars?" *The Daily Galaxy*, October 4, 2010, http://www.dailygalaxy.com/my_weblog/2010/10/the-search-for-et-should-it-focus-on-hot-stars-black-holes-and-neutron-stars-todays-most-popular.html (accessed July 2, 2011). **One frigid example is Bok globules:** Ibid.

92   **And don't forget:** Yudkowsky, Eliezer, "Artificial Intelligence as a Positive and Negative Factor in Global Risk,"

Sept. 2008, http://intelligence.org/files/AIRisk.pdf (accessed March 3, 2013).

93   **The 1986 Chernobyl:** *INSAG-7 The Chernobyl Accident: Updating of INSAG-1* (Vienna: International Atomic Energy Agency, 1992), http://www-pub.iaea.org/MTCD/publications /PDF/Pub913e_web.pdf (accessed July 2, 2011). **We have produced designs so complicated:** Charles Perrow, *Normal Accidents: Living with High-Risk Technologies* (Princeton, NJ: Princeton University Press, 1999), 11.

94   **The point of HFTs:** CBS News, "How Speed Traders Are Changing Wall Street," *60 Minutes*, October 11, 2010, http://www.cbsnews.com/stories/2010/10/07/60minutes/ main6936075.shtml (accessed July 3, 2011). **After the sale, the price:** Cohan, Peter, "The 2010 Flash Crash: What Caused It and How to Prevent the Next One," *Daily Finance*, August 18, 2010, http://www.dailyfinance.com/2010/08 /18/the-2010-flash-crash-what-caused-it-and-how-to -prevent-the-next/ (accessed July 3, 2011). **The lower price automatically:** Nanex, "Analysis of the 'Flash Crash,'" last modified June 18, 2010, http://www.nanex .net/20100506/FlashCrashAnalysis_CompleteText.html.

95   **not only unexpected:** Perrow, Charles, *Normal Accidents*, 8. **We know that a lot of algorithms:** "The Market's Black Box: Engine for Efficiency or Ever-Growing Monster?" *Paris Tech Review*, August 25, 2010, http://www .paristechreview.com/2010/08/25/market-black-box -efficiency-growing-monster/ (accessed July 2, 2011). **That monster struck again:** Matthews, Christopher, "High Frequency Trading: Wall Street Doomsday Machine?" *Time*, August 8, 2012, http://business.time.com/2012/08 /08/high-frequency-trading-wall-streets-doomsday

-machine/ (accessed September 7, 2012). **An agent which sought only:** Omohundro, "The Nature of Self-Improving Artificial Intelligence."

96   **The AI's fourth drive:** Ibid.

97   **On Omohundro's wish list:** Ibid. **With both logic and inspiration:** Ibid.

7: THE INTELLIGENCE EXPLOSION

99   **From the standpoint of existential risk:** Yudkowsky, Eliezer, "Artificial Intelligence as a Positive and Negative Factor in Global Risk," Sept. 2008, http://intelligence.org/files/AlPosNegFactor.pdf (accessed March 3, 2013).

102   **I have a quarter-baked idea:** Banks, David L., "A Conversation with I. J. Good," *Statistical Science*, vol. 11, no. 1 (1997), 1–19, http://www.web-e.stat.vt.edu/holtzman/IJGood/A_Conversation_with_IJGood_Davild_L_Banks_1992.pdf (accessed July 2, 2011).

104   **Let an ultraintelligent machine be defined:** Good, I. J., "Speculations Concerning the First Ultraintelligent Machine," in Franz L. Alt and Morris Rubinoff, eds., *Advances in Computers*, Vol. 6 (New York: Academic Press, 1965), 31–88. **The Singularity has three well-developed definitions:** Yudkowsky, Eliezer, Machine Intelligence Research Institute, "Three Major Singularity Schools," last modified September 2007, http://yudkowsky.net/singularity/schools (accessed April 2, 2010).

106   **As a boy Good's father:** Banks, David L., "A Conversation with I. J. Good." **German U-boats:** Trueman, Chris, History Learning Site, "World War Two: U-boats," last

modified 2011, http://www.historylearningsite.co.uk/u -boats.htm (accessed December 2, 2011).

107 **Each key displayed a letter:** Sales, Tony, "The Principal of the Enigma," March 2001, http://www.codesandciphers .org.uk/enigma/enigma1.htm (accessed September 5, 2011).

108 **Turing and his colleagues:** Bletchley Park National Codes Center, "Machines behind the codes," last modified 2011, http://www.bletchleypark.org.uk/content/machines .rhtm. (accessed September 6, 2011).

109 **The heroes of Bletchley Park:** Hinsley, Harry, "The Influence of ULTRA in the Second World War," Babbage Lecture Theatre, Computer Laboratory, last modified November 26, 1996, http://www.cl.cam.ac.uk/research/ security/Historical/hinsley.html (accessed September 6, 2011). **At Bletchley Turing:** Banks, "A Conversation with I. J. Good."

110 **I won't say that what Turing did:** McKittrick, David, "Jack Good: Cryptographer whose work with Alan Turing at Bletchley Park was crucial to the War effort," *The Independent*, sec. obituaries, May 14, 2009, http://www. independent.co.uk/news/obituaries/jack-good -cryptographer-whose-work-with-alan-turing-at-bletchley -park-was-crucial-to-the-war-effort-1684506.html (accessed September 5, 2011).

113 **In 1957, MIT psychologist:** McCorduck, Pamela, *Machines Who Think, A Personal Inquiry into the History and Prospects of Artificial Intelligence* (San Francisco: W. H. Freeman & Company, 1979), 87–90. **I thought neural networks:** Ibid. **The first talks:** Good based the essay on talks he gave in 1962 and 1963.

114 **Neuroscientist, cognitive scientist:** Goertzel, Ben, and
     Cassio Pennachin, eds., *Artificial General Intelligence* (Berlin/
     New York: Springer, 2007), 18.

115 **B. V. Bowden stated:** Good, "Speculations Concerning
     the First Ultraintelligent Machines."

116 **Such machines . . . could even:** Good, I. J., ed., *The
     Scientist Speculates, an Anthology of Partly Baked Ideas* (London:
     William Heinemann, Ltd. 1962.)

117 **Speculations Concerning:** Good, I. J., *The 1998 "Computer
     Pioneer Award" of the IEEE Computer Society,* Biography and
     Acceptance Speech (1998), 8.

8: THE POINT OF NO RETURN

118 **But if the technological Singularity:** Vinge, Vernor,
     "The Coming Technological Singularity," 1993, http:
     //www-rohan.sdsu.edu/faculty/vinge/misc/WER2.html.
     This quotation sounds a lot: Could Good have read
     Vinge's essay, inspired by his own earlier essay, and then
     had a change of heart? I find that unlikely. By his death
     Good had published some three million words of
     scholarship. He's the most prolific attributer I've ever read.
     And even though many of his footnotes cite his own
     papers, I believe he would have given credit to Vinge for
     his change of heart, if Vinge's essay had prompted it. Good
     would have delighted in that kind of literary recursion.

119 **It's a problem we face every time:** Vinge, Vernor, *True
     Names and Other Dangers* (Wake Forest: Baen Books, 1987), 47.

120 **Through the sixties and seventies:** Vinge, "The
     Coming Technological Singularity."

122 **Good has captured the essence of the runaway:** Ibid.

123 **Technology thinkers including:** Kelly, Kevin, "Q&A: Hacker Historian George Dyson Sits Down With Wired's Kevin Kelly," *WIRED*, February 17, 2012, http://www.wired.com/magazine/2012/02/ff_dysonqa/all/ (accessed June 5, 2012).

124 **At his home in California:** Wisegeek, "How Big is the Internet?" last modified 2012, http://www.wisegeek.com/how-big-is-the-internet.htm (accessed July 5, 2012).

130 **Per dollar spent:** Kurzweil, Ray, *The Age of Spiritual Machines* (New York: Viking Penguin, 1999), 101–105. **By about 2020 computers will be able to model the human brain:** As of 2023, no computer has been able to fully model the human brain. The human brain is incredibly complex, with over 100 billion neurons and trillions of connections between them. It is still beyond our current capabilities to fully understand and model the brain in all its complexity.

9: THE LAW OF ACCELERATING RETURNS

132 **Computing is undergoing:** King, Rachael, "IBM training computer chip to learn like a human," *SFGate.com*, November 7, 2011, http://articles.sfgate.com/2011-11-07/business/30371975_1_computers-virtual-objects-microsoft (accessed January 5, 2012).

133 **What, then, is the Singularity:** Kurzweil, Ray, *The Singularity Is Near, When Humans Transcend Biology* (New York: Viking Press, 2005), 7. **Consider J. K. Rowling's Harry Potter:** Ibid. 4.

134 **Surprisingly, however, he was indirectly responsible:** Joy, Bill, "Why the Future Doesn't Need Us," *Wired*, August 4, 1999. **But now, with the prospect of human-level computing power:** Ibid. **Book jacket flap copy** for *The Age of Spiritual Machines* (1999).

137 **And as scholar:** Grassie, William, "H-: Millennialism at the Singularity: Reflections on Metaphors, Meanings, and the Limits of Exponential Logic," *Metanexus*, August 9, 2001, http://www.metanexus.net/essay/h-millennialism -singularity-reflections-metaphors-meanings-and-limits -exponential-logic (accessed December 10, 2011).

139 **In 1971, 2,300 transistors:** Kanellos, Michael, "Moore's Law to roll on for another decade," *CNET News*, February 10, 2003, http://news.cnet.com/2100-1001-984051.html (accessed May 5, 2011).

140 **Here's a dramatic case in point:** Markoff, John, "The iPad in Your Hand: As Fast as a Supercomputer of Yore," *New York Times*, May 9, 2011, http://bits.blogs.nytimes .com/2011/05/09/the-ipad-in-your-hand-as-fast-as-a -supercomputer-of-yore/(accessed June 25, 2011). **Only Kurzweil would've been so bold:** Kurzweil, Ray, "How My Predictions Are Faring," *KurzweilAI.Net* (blog), October 2010, http://www.kurzweilai.net/predictions/download .php (accessed August 5, 2011). **By about 2020, Kurzweil estimates:** As I wrote in an earlier note, as of 2023 no computer has been able to model the human brain, so Kurzweil's 2020 estimate was incorrect.

141 **Computing speed doubles every two years:** Yudkowsky, Eliezer, "Staring into the Singularity," *Eliezer S. Yudkowsky* (blog), November 18, 1996, http://yudkowsky.net/obsolete /singularity.html (accessed September 5, 2011). **Some**

object that Moore's law will stop before 2020: In fact, Moore's law, more of a prediction than a law, still holds true in 2023. Advances in the ability to produce smaller and more efficient transistors continues to allow for increasingly powerful and efficient processors. Tardi, Carla, 2021, "Moore's Law," Investopedia, March 22, 2023, https://www.investopedia.com/terms/m/mooreslaw.asp, (accessed April 2023). 3-D processor chips developed by Switzerland's École Polytechnique Fédérale de Lausanne (EPFL): "News Mediacom," last modified January 25, 2012, http://actu.epfl.ch/news/jumpstarting-computers-with-3d-chips/ (accessed June 5, 2012).

142 Intel's new Tri-Gate transistors: Poeter, Damon, and Mark Hachman, "Next Intel Chips Will Have the World's First '3D' Transistors." *PCMAG.COM*, May 4, 2011, http://www.pcmag.com/article2/0,2817,2384897,00.asp (accessed September 5, 2011). Recently Google's cofounder Larry Page: Feeney, Lauren, "Futurist Ray Kurzweil isn't worried about climate change," *PBS.ORG Need to Know*, February 16, 2011, http://www.pbs.org/wnet/need-to-know/environment/futurist-ray-kurzweil-isnt-worried-about-climate-change/7389/ (accessed September 5, 2011).

143 We now have the actual means: Ibid.

144 Kurzweil writes that the brain has about 100: Dartmouth University computational neuroscientist Rick Granger claims each neuron in the brain is connected to many tens of thousands of other neurons. This would make the brain much faster than Kurzweil estimated in *The Age of Spiritual Machines* and *The Singularity Is Near*. If it's much faster, it's computer

equivalent in speed is farther away. But, considering LOAR, not a lot farther. **That makes about 100 trillion interneuronal connections:** Ray Kurzweil, *The Age of Spiritual Machines* (New York: Viking Penguin, 1999), 103. **The title of fastest supercomputer:** In 2023, the fastest supercomputer in the world is currently Frontier, which is located at the Oak Ridge National Laboratory in Tennessee, USA. It was announced in June 2022 and has a peak performance of 1.102 exaFLOPS, making it the first exascale supercomputer. "November 2022 TOP500," https://www.top500.org/lists/2022/11/. **But by 2005's** ***The Singularity Is Near:*** Kurzweil, Ray, *The Singularity Is Near* (New York: Viking Press, 2005), 71.

145 **Perhaps as Kurzweil says:** Kurzweil, Ray, "Response to Mitchell Kapor's 'Why I Think I Will Win,'" *KurzweilAI. net*, April 20, 2002, http://www.kurzweilai.net/response -to-mitchell-kapor-s-why-i-think-i-will-win (accessed September 5, 2011). **That means writing more complex algorithms:** Shulman, Carl, and Anders Sandberg, Machine Intelligence Research Institute, "Implications of a Software-Limited Singularity," last modified October 31, 2010, http://intelligence.org/files/softwarelimited.pdf (accessed March 3, 2013). **Faster computers contribute:** Ibid. **more useful tools:** Know what else doubles about every two years? The Internet, and all the components that make it faster, more densely connected, and able to assimilate more data. In 2009 Google estimated that the Internet contained about five million terabytes of data— that's 250,000 times more information than all the books in the Library of Congress. By 2011 it will contain about 500,000 times more data than the LoC. Harris Interactive,

an Internet-based market research and polling firm, announced that growth in the number of Internet *users* justifies its description as the "fastest growing technology in history." Four years ago, in 2008, the Internet had just under 1.2 billion users worldwide. In 2010 there were more than two billion. **Kurzweil writes that:** Kurzweil, *The Singularity Is Near*, 2005. **That means plugging:** National Institute on Deafness and Other Communication Disorders, "More About Cochlear Implants," last modified June 7, 2010, http://www.nidcd.nih.gov/health/hearing /pages/coch_moreon.aspx (accessed September 15, 2011).

10: THE SINGULARITARIAN

148   **In contrast with our intellect:** McAuliffe, Wendy, "Hawking warns of AI world takover," *ZDNet*, September 3, 2001, http://www.zdnet.co.uk/news/application-development /2001/09/03/hawking-warns-of-ai-world-takeover -2094424/ (accessed September 5, 2011). **Within thirty years, we will have the technological means:** Vinge, Vernor, "The Coming Technological Singularity," 2003, http://www-rohan.sdsu.edu/faculty/vinge/misc /WER2.html (accessed September 5, 2011).

151   **If the consequences of an action:** Kurzweil, Ray, *The Singularity Is Near* (New York: Viking Press, 2005), 403.

153   **As Steve Omohundro warns:** Omohundro, Stephen, "The Basic AI Drives," November 11, 2007, http://self awaresystems.com. **it may have other uses for our atoms:** Yudkowsky, Eliezer, "Artificial Intelligence as a Positive and Negative Factor in Global Risk," August 31,

2006,   http://intelligence.org/files/AIPosNegFactor.pdf
(accessed February 28, 2013).

154   **We can't just say, "we'll put in this little software code":** Kurzweil, Ray, "Ray Kurzweil: The H+ Interview," *H+ Magazine*, December 30, 2009, http://hplusmagazine .com/2009/12/30/ray-kurzweil-h-interview/ (accessed March 1, 2011). **our most sensitive systems, including aircraft avionics:** Ukman, Jason, and Ellen Nakashima, "24,000 Pentagon files stolen in major cyber breach official says," *Washington Post*, sec. national, July 14, 2011, http://www.washingtonpost.com/blogs/checkpoint -washington/post/24000-pentagon-files-stolen-in-major -cyber-breach-official-says/2011/07/14/gIQAsaaVEI_blog .html (accessed September 28, 2011).

155   **There's a lot of talk about existential risk:** Kurzweil, Ray, "Ray Kurzweil: The H+ Interview."

157   **exhibit unethical decision-making tendencies:** Kiff, Paul, Daniel Stancato, Stephane Cote, Rodolfo Mendoza-Denton, and Dacher Keltner, "Higher social class predicts increased unethical behavior," *Proceedings of the National Academy of Sciences*, no. 26 (January 2012), http: // www.pnas.org/content/early/2012/02/21/1118373109 .abstract (accessed February 11, 2012).

158   **The Singularitarians' conceit:** Incidentally, there's some interesting writing on the Web about the concept of a Singleton. Conceived by ethicist Nick Bostrom, a "Singleton" is a sole dominant AI in control of decisions at the highest level. See Bostrom, Nick, "What is a Singleton?" last modified 2005, http://www.nickbostrom .com/fut/singleton.html (accessed September 19, 2011).

159   **Technologists are providing almost religious visions:**

Markoff, John, "Scientists Worry Machines May Outsmart Man," *New York Times*, sec. science, July 25, 2009, http://www.nytimes.com/2009/07/26/science/26robot.html (accessed September 25, 2011). **costly, unforeseen behaviors:** Horvitz, John, AAAI, "Interim Report from the Panel Chairs AAAI Presidential Panel on Long-Term AI Futures August 2009," last modified 2009, http://research.microsoft.com/en-us/um/people/horvitz/note_from _AAAI_panel_chairs.pdf (accessed September 28, 2011).

160   **I went in very optimistic about the future of AI:** Markoff, John, "Scientists Worry Machines May Outsmart Man."

10: A HARD TAKEOFF

163   **However, an intelligence explosion may be unavoidable:** I'm not sure this is true for Marcus Hutter's AIXI, though experts tell me it is. But since AIXI is uncomputable, it would never be a candidate for an intelligence explosion anyway. AIXItl—a computable approximation of AIXI—is another matter. This is also probably not true of mind uploading, if such a thing ever comes to pass.

164   **Computer science-based researchers want to engineer AGI:** The mind versus brain debate is too large to address here.

167   **with $50 million in grants:** Lenat, Doug, "Doug Lenat on Cyc, a truly semantic Web, and artificial intelligence (AI)," *developerWorks*, September 16, 2008, http://www.ibm.com/developerworks/podcast/dwi/cm-int091608txt .html (accessed September 28, 2011). **Carnegie Mellon**

University's NELL: Lohr, Steve, "Aiming to Learn as We Do, a Machine Teaches Itself," *New York Times*, sec. science, October 4, 2010, http://www.nytimes.com/2010/10/05 /science/05compute.html?pagewanted=all (accessed September 28, 2011).

169 **Many, especially those at MIRI:** Some Singularitarians want to get to AGI as soon as possible, owing to its potential to alleviate human suffering. This is Ray Kurzweil's position. Others feel achieving AGI will move them closer to ensuring their own immortality. MIRI's founders, including Eliezer Yudkowsky, hope AGI takes a long time because the likelihood of destroying ourselves may diminish with time due to more, better research. **these drives are: efficiency, self-preservation, resource acquisition, and creativity:** Omohundro, Stephen, "The Basic AI Drives," November 11, 2007, http://self awaresystems.com/2007/11/30/paper-on-the-basic-ai -drives/ (accessed June 1, 2011)

171 **iRobot, the company he founded, manufactures weaponized robots:** Palmisano, John, "iRobot Demonstrates New Weaponized Robot," *IEEE Spectrum*, May 30, 2010, http://spectrum.ieee.org/automaton /robotics/military-robots/irobot-demonstrates-their -latest-war-robot (accessed October 2, 2011).

172 **an intelligence explosion requires AGI:** Loosemore, Richard, and Goertzel, Ben, "Why an Intelligence Explosion is Probable," *H+ Magazine*, March 7, 2011, http://hplusmagazine.com/2011/03/07/why-an-intelligence -explosion-is-probable/ (accessed October 2, 2011). **self- improvement in pursuit of goals is rational behavior:** Omohundro, Stephen, "The Nature of Self-Improving

Artificial Intelligence," January 21, 2008, http://
selfawaresystems.files.wordpress.com/2008/01/nature
_of_self_improving_ai.pdf (accessed September 4, 2010).

173 **Goertzel's plan is to create an infantlike AI "agent":**
Dvorsky, George, "How will we build an artificial human
brain?" *io9 We Come from the Future*, May 2, 2012, http://
io9.com/5906945/how-will-we-build-an-artificial-human
-brain (accessed June 2, 2012). **Prevent an intelligence
explosion from occurring in this virtual world:** Hutter,
Marcus, "Can Intelligence Explode?" February 2012,
singularitysummit.com.au/2012/08/can-intelligence-
explode (accessed July 3, 2012)

174 **software that derives scientific laws from raw data:**
Kelm, Brandon, "Download Your Own Robot Scientist,"
Wired, December 3, 2009, http://www.wired.com/
wiredscience/2009/12/download-robot-scientist/
(accessed June 3, 2011). **many generations later output
physical laws:** Chang, Kenneth, "Hal, Call Your Office:
Computers That Act Like Physicists," *New York Times*, April 2,
2009, http://www.nytimes.com/2009/04/07/science/07robot
.html?em (accessed July 5, 2012). **it evolved rules about
its own operation:** Johnson, George, the Alicia Patterson
Foundation, "Eurisko, the Computer with a Mind of Its
Own," last modified April 6, 2011, http://aliciapatterson.org
/stories/eurisko-computer-mind-its-own (accessed July 5,
2012). **Eurisko's greatest success:** Ibid.

175 **Eurisko created a rule:** Lenat, Douglas B., "EURISKO:
A Program That Learns New Heuristics and Domain
Concepts (The Nature of Heuristics III: Program Design
and Results)," *Artificial Intelligence*, 21 (1983), 61–98.

176 **he urges programmers not to bring it back:** Yudkowsky,

Eliezer, "Let's reimplement EURISKO!" *Less Wrong* (blog), June 11, 2009, http://lesswrong.com/lw/10g/lets_reimplement_eurisko/ (accessed June 3, 2010).

177 **Every eight processors of its 30,000:** Mozyblog, "How Much is a Petabyte?" last modified July 2, 2009, http://mozy.com/blog/misc/how-much-is-a-petabyte/ (accessed April 3, 2010). **as a company spokesman put it:** Brodkin, John, "$1,279-per-hour, 30,000-core cluster built on Amazon EC2 cloud," *Ars Technica,* last modified September 21, 2011, http://arstechnica.com/business/news/2011/09/30000-core-cluster-built-on-amazon-ec2-cloud.ars (accessed April 3, 2012).

179 **She selects jobs to offer a sailor:** Franklin, Stan, and F. G. Patterson, "The Lida Architecture: Adding New Modes of Learning to an Intelligent, Autonomous Software Agent," Institute for Intelligent Systems, FedEx Institute of Technology, The University of Memphis, June 2006, http://www.theassc.org/files/assc/zo-1010-lida-060403.pdf (accessed February 23, 2010). **We are forming Silicon Valley's next great company:** Stealth-Company, "Get In Early," last modified 2008, http://www.stealth-company.com/site/home (accessed December 2, 2011).

180 **DARPA funded more AI research than private corporations:** *Funding a Revolution: Government Support for Computing Research* (Washington, D.C: National Academy Press, 1999), 200–205. **How is DARPA spending its money?:** *Department of Defense Fiscal Year (FY) 2012 Budget Estimates Defense Advanced Research Projects Agency* (Arlington, Virginia: DOD, 2011).

181 **The Cognitive Computing Systems program:** Ibid.

182 **to create cognitive software systems:** SRI International,

"Cognitive Assistant that Learns and Organizes," last modified 2012, http://www.ai.sri.com/project/CALO. (accessed March 2, 2010). **Within its own cognitive architecture:** Ibid.

183   **In 2008, Apple Computer bought Siri:** Schonfeld, Erick, "Silicon Valley Buzz: Apple Paid More Than $200 Million for Siri to Get into Mobile Search," *TechCrunch*, last modified April 28, 2010, http://techcrunch.com/2010 /04/28/apple-siri-200-million/ (accessed March 10, 2011). **the Army will take iPhones into battle:** Raice, Shayndi, "Smartphones Going into Battle, Army Says," *Digits: Technology News and Insights* (blog), December 14, 2010, http://blogs.wsj.com/digits/2010/12/14/smartphones-going -into-battle-army-says/ (accessed March 10, 2010).

184   **stunning implications for the world economy:** Loose-more, Richard, and Ben Goertzel, "Why an Intelligence Explosion is Probable," *H+ Magazine*, March 7, 2011, http: //hplusmagazine.com/2011/03/07/why-an-intelligence -explosion-is-probable/ (accessed April 2, 2011). **any limitations to the economic growth rate:** Ibid.

185   **growth rate would be defined by the various AGI projects:** Ibid.

12: THE LAST COMPLICATION

187   **How can we be so confident:** Hibbard, Bill, "AI is a Threat Despite Calming Voices," last modified August 20, 2010, http://sites.google.com/site/whibbard/g/hibbard_oped _aug2010. (accessed June 10, 2011). **we will eventually uncover the principles:** Anissimov, Michael, *Accelerating*

Future, "More Singularity Curmudgeonry from John Horgan," last modified June 23, 2010 (accessed June 11, 2011).

188 **Normalcy Bias:** Valentine, Pamela, and Thomas Smith, "Finding Something to Do: The Disaster Continuity Care Model," *Brief Treatment and Crisis Intervention*, 2, (Summer 2002), 183–196, http://btci.edina.clockss.org/cgi/reprint /2/2/183.pdf (accessed September 4, 2012).

189 **The same restriction would apply to human-computer interfaces:** As we look into the software complexity problem, let's also consider how long humans have been trying to scratch this particular itch. In 1956, John McCarthy, called the "father" of artificial intelligence (he coined the term) claimed the whole problem of AGI could be solved in six months. In 1970, AI pioneer Marvin Minsky said, "In from three to eight years we will have a machine with the general intelligence of an average human being." Considering the state of the science, and with the benefits of hindsight, both men were loaded with hubris, in the Classical sense. Hubris comes from a Greek word meaning arrogance, and often, arrogance toward the gods. The sin of hubris was attributed to men who tried to act outside of human limitations. Think Icarus attempting flight, Sisyphus outwitting Zeus (for a while anyway), and Prometheus giving fire to man. Pygmalion, according to mythology, was a sculptor who fell in love with one of his statues, Galatea, Greek for "sleeping love." Yet he suffered no cosmic comeuppance. Instead, Aphrodite, Goddess of Love, brought Galatea to life. Hephaestus, Greek God of technology, among other things, routinely built metal automatons to help with his

metallurgy. He created Pandora *and* her box, and Talos, a giant made of bronze that protected Crete from pirates.

Paracelsus, the great medieval alchemist, best known for linking medicine and chemistry, allegedly fine-tuned a formula for creating humanlike creatures, and human-animal hybrids, called homunculi. Just fill a bag with human bones, hair, and sperm, then bury it in a hole along with some horse manure. Wait forty days. A humanlike infant will struggle to life, and thrive if fed on blood. It will always be tiny but it will do your bidding until it turns on you and runs away. If you'd like to mix the human with another animal, say a horse, substitute horsehair for human hair. However, while I can think of ten uses for a tiny human (cleaning heating ducts, getting dog hair out of the Roomba, and more), I can think of none for a tiny centaur.

Before MIT's Robotics Lab and Mary Shelley's *Frankenstein* existed, there was the Jewish tradition of the golem. Like Adam, a golem is a male creature made of earth. Unlike Adam, it was not brought to life by the breath of God, but by incantations of words and numbers uttered by rabbis (Kabbalists who believe in an orderly universe and the divinity of numbers). The name of God, written on paper and put in its mouth, kept the mute, ever-growing creature "alive." In Jewish folklore, magic-wielding rabbis created golems to serve as valets and domestic servants. The most famous golem, named Yosele, or Joseph, was created in the sixteenth century by Prague's chief rabbi, Yehuda Loew. In an era when Jews were accused of using the blood of Christian infants to make matzoth, Yosele kept busy rounding up gentile "blood" libelers, capturing crooks in

Prague's ghetto, and generally helping Rabbi Loew fight crime. Eventually, according to tradition, Yosele went berserk. To save his fellow Jews, the rabbi battled the golem, and removed the life-giving piece of paper from his mouth. Yosele turned back to clay. In another version, Rabbi Loew was crushed to death by the falling giant, a fitting reward for the hubristic act of creation. In yet another version, Rabbi Loew's wife asked Yosele to fetch water. He kept at it until his creator's house was flooded. In computer science, not knowing whether or not your program will act this way is called the "halting problem." That is, good programs will run until their instructions tell them to stop, and in general it is impossible to know for sure if any given program will ever stop until you run it. Taken here, Rabbi Loew's wife could have specified *how much* water to fetch, say, one hundred liters, and Yosele probably would've stopped after that. In this story she neglected to.

The halting problem is a real issue to programmers, who may not discover until their programs are running that an infinite loop lies hidden in the code. And an interesting thing about the halting problem is that it's impossible to create a program that determines if the program you've written *has* the halting problem. That diagnostic debugger sounds plausible, but none other than Alan Turing discovered it is not (and he discovered it before there were computers *or* programming). He said the halting problem is unsolvable because if the debugger encounters a halting problem in the target program, it will succumb to the infinite loop while analyzing it, and never determine if the halting problem was there. You, the programmer, would be waiting for it to come up with the answer for the same amount of time you'd wait for the original program to halt. That is, a very long time, perhaps forever. Marvin Minsky, one of the fathers of artificial intelligence, pointed out that "any finite-state machine, if left completely to itself, will fall eventually into a perfectly periodic repetitive pattern. The duration of this repeating pattern cannot exceed the number of internal states of the machine." Translated, that means a computer of average memory, while running a program with a halting problem, would take a very long time to fall into a pattern of repetition, which could then be detected by a diagnostic program. How long? Longer than the universe will exist, for some programs. So, for *practical* purposes, the halting problem means it is impossible to know whether any given program will finish.

Once Rabbi Loew noticed Yosele's inability to stop, he could have fixed it with a patch (a change to its programming), in this case by taking the paper out of the giant's mouth on which was written the name of God. In the end, Yosele was shut down and stored, it is said, in the attic of the Old New Synagogue in Prague, to come alive again at the end of days. Rabbi Loew, an actual, historical character, is buried in Prague's Jewish Cemetery (fittingly, not so far from Franz Kafka's grave). So alive is the myth of Yosele to families of Eastern European Jewish descent that as late as the last century children were taught the rhyme that will awaken the golem in the end times.

Rabbi Loew's fingerprints are all over the cultural descendents of the golem, from the obvious *Frankenstein*, to J.R.R. Tolkien's *The Lord of the Rings*, to the Hal 9000 computer of Stanley Kubrick's classic movie *2001: A Space Odyssey*. The computer science experts Kubrick recruited to advise him about the homicidal robot included Marvin Minsky and I. J. Good. Good had only recently written about the intelligence explosion, and anticipated it was within two decades away. Probably to his bemusement, advising Kubrick about Hal led to Good's 1995 induction into the Academy of Motion Picture Arts and Sciences.

According to the history of AI author Pamela McCorduck, her interviews revealed that a handful of the pioneers of computer science and artificial intelligence believe that they were directly descended from Rabbi Loew. They include John von Neumann and Marvin Minsky.

190   **Entrepreneur and AI maker Peter Voss:** Voss, Peter, *MIRI Interview Series*, 2011, http://citationmachine.net/index2.php?reqstyleid=10&mode=form&rsid=&reqsrcid=ChicagoInterview&more=yes&nameCnt=1 (accessed June 10, 2010). **Google's proprietary algorithm called PageRank:** Geordie, "Learn How Google Works: in Gory Detail," *PPC Blog* (blog), 2011, http://ppcblog.com/how-google-works/ (accessed October 10, 2011).

192   **mankind's primary tool:** Schwartz, Evan, "The Mobile Device is Becoming Humankind's Primary Tool," *Technology Review*, November 29, 2010, http://www.technologyreview.com/news/421826/the-mobile-device-is-becoming-humankinds-primary-tool-infographics-feature/ (accessed December 4, 2011).

193   **you merely think of a question:** Carr, Nicholas, "When Google Grows Up," *Forbes.com*, January 11, 2008, http://www.forbes.com/2008/01/11/google-carr-computing-tech-enter-cx_ag_0111computing.html (accessed March

10, 2011). **You are never lost:** Kharif, Olga, "Google Uses AI to Make Search Smarter," *Bloomberg Businessweek*, September 21, 2010, http://www.businessweek.com/stories /2010-09-21/google-uses-ai-to-make-search-smarter businessweek-business-news-stock-market-and-financial -advice (accessed April 5, 2012). **Siri will interact with online retailers:** Li, Wendi, "Improved Siri Will Do Everything for You, Including Shopping: Apple Patent Filing," *International Business Times*, January 21, 2012, http://www.ibtimes.com/articles/285440/20120121/siri -shopping-apple-patent-filing-ipad-3.htm (accessed March 10, 2012).

194 **Andrew Rubin, Google's Senior Vice President of Mobile:** Fried, Ina, "Android Chief Says Your Phone Should Not Be Your Assistant," *All Things D*, October 19, 2011, http://allthingsd.com/20111019/android-chief-says-your -phone-should-not-be-your-assistant/ (accessed November 13, 2011).

198 **It may be that we need a scientific breakthrough:** Goertzel, Ben, "Editor's Blog Report on the Fourth Conference on Artificial General Intelligence," *H+ Magazine*, September 1, 2011, http://hplusmagazine.com/2011/09/01/ report-on-the-fourth-conference-on-artificial-general -intelligence/ (accessed November 22, 2011). **LIDA scores like a human:** Biever, Celeste, "Bot shows signs of consciousness," *New Scientist*, April 1, 2011, http://www .newscientist.com/article/mg21028063.400-bot-shows -signs-of-consciousness.html (accessed June 1, 2011).

200 **committing the Holocaust:** Goertzel, Ben, "The Machine Intelligence Research Institute's Scary Idea (and Why I Don't Buy It)," *The Multiverse According to Ben*

(blog), October 29, 2010, http://multiverseaccordingtoben. blogspot.com/2010/10/singularity-institutes-scary-idea-and.html (accessed June 1, 2011).

201 **Converting an AI system to AGI through brute force:** Loosemore, Richard, and Ben Goertzel, "Why an Intelligence Explosion Is Probable," *H+ Magazine,* March 7, 2011, http://hplusmagazine.com/2011/03/07/why-an-intelligence-explosion-is-probable/ (accessed November 25, 2011).

203 **A lot of cutting edge AI:** "AI set to exceed human brain power," *CNN Tech,* July 24, 2006, http://articles.cnn.com/2006-07-24/tech/ai.bostrom_1_neural-networks-human-brain-turing-test?_s=PM:TECH (accessed November 25, 2011).

204 **easy to make computers exhibit adult-level performance:** Moravec, Hans, *Mind Children* (Cambridge: Harvard University Press, 1988), 15.

205 **tell the difference between a dog and a cat:** Though this is about to change, thanks to Dartmouth's Richard Granger, profiled in this chapter. **reasoning is much easier than perceiving:** Moravec, Hans, Robotics Institute, Carnegie Mellon University, "The Age of Robots," June 1993, http://www.frc.ri.cmu.edu/~hpm/project.archive/general.articles/1993/Robot93.html (accessed March 19, 2011).

206 **formalization reveals hidden rules:** Granger, Richard, "How Brains Are Built: Principles of Computational Neuroscience," *Cerebrum* (January 2011), http://www.dana.org/news/cerebrum/detail.aspx?id=30356 (accessed June 3, 2011).

207 **Every structure has been precisely shaped:** Allen, Paul, and Mark Greaves, "Paul Allen: The Singularity

Isn't Near," *Technology Review*, November 12, 2011, http://www.technologyreview.com/blog/guest/27206/ (accessed November 25, 2011).

208 **Our goal in computational neuroscience:** Granger, "How Brains Are Built: Principles of Computational Neuroscience."

209 **Intelligence will also win the day:** As Granger writes in his book, *Big Brains*, Neanderthals had larger brains than we do, and might have been more intelligent. However, that they were more intelligent isn't by any means certain.

210 **what indication of its existence might we expect:** Dyson, George, *Edge*, "Turing's Cathedral," last modified October 24, 2005, http://www.edge.org/conversation turing-395-cathedral (accessed April 20, 2011).

13: UNKNOWABLE BY NATURE

211 **Both because of its superior planning ability:** Bostrom, Nick, Oxford University, "Ethical Issues in Advanced Artificial Intelligence," last modified 2003, http://www.nickbostrom.com/ethics/ai.html (accessed April 24, 2011).

218 **Basically, we are looking for:** Kurzweil, Ray, "Kurzweil Responds: Don't Underestimate the Singularity," *Technology Review*, October 19, 2011, http://www.technologyreview.com/view/425818/kurzweil-responds-dont-underestimate-the/ (accessed November 1, 2011).

219 **to get Watson to understand what people say:** Nuance Communications, Inc, "IBM to Collaborate with Nuance to Apply IBM's Watson Analytics Technology to Healthcare,"

last modified February 17, 2011, http://www.nuance.com
/company/news-room/press-releases/NC_008477
(accessed June 18, 2011).

220 **Its hardware is massively parallel:** "What is Watson?"
IBM, 2011, http://www-03.ibm.com/innovation/us/watson
(accessed August 18, 2011). **parallelism can handle
staggering computational workloads:** Ferrucci, David,
Eric Brown, Jennifer Chu-Carroll, James Fan, David
Gondek, Aditya Kalyanpur, Adam Lally, William Murdock,
Eric Nyberg, John Prager, Nico Schlaefer, and Chris Welty,
"Building Watson: An Overview of the DeepQA Project,"
AI Magazine (Fall 2010), http://www.aaai.org/ojs/index
.php/aimagazine/article/view/2303 (accessed August 18,
2011).

222 **A lot has been written:** Kurzweil, Ray, "Kurzweil
Responds: Don't Underestimate the Singularity."

223 **Searle said:** Blumenthal, Andy, "Watson Can Swim," *The
Total CIO* (blog), March 14, 2011, http://andyblumenthal.
posterous.com/watson-can-swim (accessed May 1, 2011).
Can a submarine swim?: Ibid. **a submarine doesn't
"swim":** Ibid. **The computer's techniques:** Jennings,
Ken, "My Puny Human Brain," *Slate*, February 26, 2011,
http://www.slate.com/articles/arts/culturebox/2011/02
/my_puny_human_brain.single.html (accessed May 22,
2011).

225 **Ohio State (17) and Kansas (14):** "Ohio State, Kansas,
BYU headline poll," *ESPN Men's Basketball*, March 1, 2011,
http://sports.espn.go.com/ncb/news/story?id=6167338
(accessed January 18, 2012).

227 **people who have become paraplegics:** Solomon,
Robert C., *Thinking About Feeling, Contemporary Philoso-*

phers on Emotions (New York: Oxford University Press, 2004), 47, 48 (accessed January 21, 2012).

14: THE END OF THE HUMAN ERA?

229 **The argument is basically very simple:** Perrow, Charles, *Normal Accidents: Living with High-Risk Technologies* (New York: Basic Books, 1984), 4. **we are just a few years away from a major catastrophe:** Anissimov, Michael, "The Road to the Singularity," *Accelerating Future* (blog), November 19, 2007, http://www.acceleratingfuture.com/people-blog/2007/the-road-to-the-singularity/ (accessed March 10, 2011).

230 **Thus, if we evolve a complex system:** Jurvetson, Steve, "The Dichotomy of Design and Evolution," *The J Curve* (blog), July 13, 2006, http://jurvetson.blogspot.com/2006/07/dichotomy-of-design-and-evolution.html (accessed October 10, 2011).

232 **a few "minor" accidents would be desirable:** Whitby, Blay, *Reflections on Artificial Intelligence* (Exeter: Intellect Ltd., 1996), 31. **it learns based on the right answers:** Ferrucci, David, "A: This Computer Could Defeat You at 'Jeopardy!' Q: What is Watson?" February 14, 2011, http://www.pbs.org/newshour/bb/science/jan-june11/jeopardy_02-14.html (accessed October 10, 2011).

233 **it will follow its own drives:** Omohundro, Stephen, "The Basic AI Drives," November 11, 2007, http://selfawaresystems.com/2007/11/30/paper-on-the-basic-ai-drives/ (accessed June 21, 2011).

234 **none, except Omohundro:** Relative to scientists engaged in the pursuit, Yudkowsky and MIRI are not trying to

create AGI, though they consider the ethics of creating it and how to control it. AGI maker Ben Goertzel has frequently written about AI ethics, but that's not the same as focusing on solutions to AI dangers.

237 **The scientists at Asilomar:** Barinaga, Marcia, "Asilomar Revisited: Lessons for Today?" *Science*, March 3, 2000, http://www.biotech-info.net/asilomar_revisited.html (accessed October 10, 2011). **10 percent of the world's cropland:** International Service for the Acquisition of Agri-Biotech Applications, "Crop Biotech Update," last modified February 22, 2011, http://www.isaaa.org/kc /cropbiotechupdate/specialedition/2011/default.asp (accessed October 10, 2011).

238 **programmed to *die by default*:** Sterrit, Roy, *Apoptotic Robotics Programmed Death by Default*, "2011 Eighth IEEE International Conference and Workshops on Engineering of Autonomic and Autonomous Systems," last modified February 11, 2011, http://ieeexplore.ieee.org/xpl/freeabs _all.jsp?arnumber=5946191 (accessed October 10, 2011). **Every time a cell divides:** Ibid.

239 **all computer-based systems should be apoptotic:** Ibid.

240 **Called the "Safe-AI Scaffolding Approach":** Omohundro, Stephen, Self-Aware Systems, "Rational Artificial Intelligence for the Greater Good," last modified March 30, 2012, http://selfawaresystems.com/2012/03 /30/rational-artificial-intelligence-for-the-greater-good/ (accessed July 10, 2012). **Given the infrastructure:** From a September 6, 2008, correspondence between Stephen Omohundro and Eric Baum. **powerful enough to address all the problems:** Ibid.

242 **greater intelligence will always find a way:**

Kurzweil, Ray, *The Singularity Is Near* (New York: Viking Press, 2005), 424.

15: THE CYBER ECOSYSTEM

244 **The next war will begin in cyberspace:** Lopez, C. Todd, WWW.ARMY.MILL, "Next War Will Begin in Cyberspace Experts Predict," last modified February 27, 2009, http://www.army.mil/article/17561/Next_war_will_begin_in_cyberspace__experts_predict/ (accessed October 10, 2011). **I am selling a private zeus:** PHPSeller, OpenSC.ws, "Malware Samples and Information Forum," last modified August 2009, http://www.opensc.ws/malware-samples-information/7862-sale-zeus-1-2-5-1-clean.html (accessed October 10, 2011).

245 **the Internet's immune system:** Kopytoff, Verne, "Deploying New Tools to Stop the Hackers," *New York Times*, sec. technology, June 17, 2011, http://www.nytimes.com/2011/06/18/technology/18security.html?pagewanted=all (accessed October 10, 2011). **malware passed good software:** Ibid.

246 **Anonymous has attacked the Vatican:** Reuters, "Hackers group Anonymous takes down Vatican website," *Huffington Post*, July 7, 2012, http://www.huffingtonpost.com/2012/03/07/anonymous-hacks-vatican-website_n_1327297.html (accessed July 11, 2012). **In 2011, botnet victims increased 654 percent:** Schwartz, Mathew, "Botnet Victims Increased 654 percent in 2011," *Information Week*, February 18, 2011, http://www.informationweek.com/news/security/attacks/229218944?cid=RSSfeed

_IWK_All (accessed July 11, 2012). **a one trillion-dollar industry:** Symantec, "What is Cybercrime?" last modified 2012, http://us.norton.com/cybercrime/definition .jsp (accessed July 11, 2012).

247 **Cloud computing has been a runaway success:** Malik, Om, "How Big is Amazon's Cloud Computing Business? Find Out," *GIGAOM*, August 11, 2010, http://gigaom.com/cloud /amazon-web-services-revenues/ (accessed June 4, 2011). **Zeus stole some $70 million:** Ragan, Steve, "ZBot data dump discovered with over 74,000 FTP credentials," *The Tech Herald*, June 29, 2009, http://www.thetechherald.com /articles/ZBot-data-dump-discovered-with-over-74-000 -FTP-credentials/6514/ (accessed June 4, 2011).

249 **21.3 percent overall, comes from Shaoxing:** Melanson, Donald, "Symantec names Shaoxing, China, as world's malware capital," *Engadget*, March 29, 2010, http://www. engadget.com/2010/03/29/symantec-names-shaoxing -china-worlds-malware-capital (accessed June 4, 2011).

250 **cybertheft helps support China's economy:** Gross, Michael Joseph, "Enter the Cyber-dragon." *Vanity Fair*, September 2011, http://www.vanityfair.com/culture/ features/2011/09/chinese-hacking-201109 (accessed May 1, 2012). **Why spend $300 billion:** Gorman, Siobhan, August Cole, and Yochi Dreazen, "Computer Spies Breach Fighter-Jet Project," *Wall Street Journal*, sec. technology, August 21, 2009, http://online.wsj.com/article/SB1240274 91029837401.html (accessed May 1, 2012). **From 2007 to 2009 an average of 47,000:** Sterner, Eric, "Retaliatory Deterrence in Cyberspace," *Strategic Studies Quarterly* (Spring 2011), http://www.marshall.org/article.php ?id=933 (accessed May 1, 2012).

251 **the right under the law of armed conflict:** Lynn III, William, "The Pentagon's Cyberstrategy, One Year Later," *Foreign Affairs*, September 28, 2011.

252 **Some three thousand organizations:** "The Future of the Electric Grid," *MIT ENERGY INITIATIVE*, 2011, http://web.mit.edu/mitei/research/studies/the-electric-grid-2011.shtml (accessed May 1, 2012).

253 **The highly interconnected grid:** Ibid.

254 **analysis of hypothetical disasters:** "Terrorism and the Emp Threat to Homeland Security." *Hearing Before the Subcommittee on Terrorism, Technology and Homeland Security of the Committee on the Judiciary United States Senate One Hundred Ninth Congress First Session*, March 8, 2005, http://www.gpo.gov/fdsys/pkg/CHRG-109shrg21324/pdf/CHRG-109shrg21324.pdf (accessed March 1, 2010). **Significant disruptions in any one of these sectors:** Lynn III, William, United States Department of Defense, "Remarks on the Department of Defense Cyber Strategy," last modified July 14, 2011, http://www.defense.gov/speeches/speech.aspx?speechid=1593 (accessed February 8, 2012).

255 **turn out the lights for fifty million people:** McAfee and CSIS, "In the Dark: Crucial Industries Confront Cyberattacks," last modified 2011, http://www.mcafee.com/us/resources/reports/rp-critical-infrastructure-protection.pdf (accessed February 8, 2012). **Many industrial generators and transformers:** USDOE and NERC, "High-Impact, Low-Frequency Event Risk to the North American Bulk Power System," last modified June 2010, http://www.nerc.com/files/HILF.pdf (accessed February 8, 2012).

256 **the generator's fans grew steadily louder:** Philipp, Joshua, "Critical Infrastructure Vulnerable in Cyber-Attacks," *The Epoch Times*, May 13, 2011, http://www.theepochtimes.com/n2/technology/critical-infrastructure-vulnerable-in-cyber-attacks-56273.html (accessed February 10, 2012). **The device that controlled DHS' tortured generator:** Associated Press, "US video shows hacker hit on power grid," *China Daily*, September 27, 2007, http://www.chinadaily.com.cn/world/2007-09/27/content_6139437.htm (accessed February 10, 2012).

257 **it was built to kill industrial machines:** Bres, Eric, "The Stuxnet Mystery Continues," *Tofino* (blog), October 10, 2010, http://www.tofinosecurity.com/blog/stuxnet-mystery-continues (accessed June 14, 2012). **holes that permit unauthorized access:** IT Networks, "Stuxnet Things You Don't Know," last modified March 25, 2011, http://www.it-networks.org/2011/03/25/stuxnet-things-you-dont-know/ (accessed December 14, 2011).

258 **their operators didn't sense anything wrong:** Poeter, Damon, "Former NSA Head: Hitting Iran with Stuxnet Was a 'Good Idea,'" *PCMAG.COM*, March 12, 2012, http://www.pcmag.com/article2/0,2817,2401111,00.asp (accessed April 22, 2012). **two countries jointly created Stuxnet:** Ibid. **a joint U.S.-Israel cyberwar campaign against Iran:** Sanger, David, "Obama Order Sped Up Wave of Cyberattacks Against Iran," *New York Times*, June 1, 2012, http://www.nytimes.com/2012/06/01/world/middleeast/obama-ordered-wave-of-cyberattacks-against-iran.html?_ (accessed June 14, 2012).

259 **Duqu and Flame are reconnaissance viruses:** "W32.

Duqu: The Precursor to the Next Stuxnet," *Symantec Connect* (blog), October 24, 2011, http://www.symantec.com/connect/w32_duqu_precursor_next_stuxnet (accessed January 14, 2012).

260 **[Stuxnet's creators] opened up the box:** Sean McGurk, former head of cybersecurity DHS, interview by Steve Kroft, "Stuxnet: Computer worm opens new era of warfare," CBS News, March 4, 2012, http://www.cbsnews.com/8301-18560_162-57390124/stuxnet-computer-worm-opens-new-era-of-warfare/ (accessed June 3, 2012).

261 **Before, a Stuxnet-type attack:** Clayton, Mark, "From the man who discovered Stuxnet, dire warnings one year later," *MinnPost*, September 23, 2011, http://www.minnpost.com/christian-science-monitor/2011/09/man-who-discovered-stuxnet-dire-warnings-one-year-later (accessed January 14, 2012). **the good luck did not last:** Sanger (2012).

263 **Al Qaeda's attacks of 9/11:** Carter, Shan, and Amanda Cox, "One 9/11 Tally: $3.3 Trillion," *New York Times*, September 8, 2011, http://www.nytimes.com/interactive/2011/09/08/us/sept-11-reckoning/cost-graphic.html (accessed January 14, 2012). **The subprime mortgage scandal:** International Monetary Fund, "IMF Loss Estimates: Executive Summary," last modified 2010, http://www.imf.org/external/pubs/ft/weo/2009/01/pdf/exesum.pdf (accessed October 13, 2011). **The Enron Scandal comes in at about $71 billion:** Laws.Com, "Easy Guide to Understanding ENRON," last modified December 6, 2011, http://finance.laws.com/enron-scandal-summary (accessed January 14, 2012). **while the Bernie Madoff**

fraud: Graybow, Martha, "Madoff mysteries remain as he nears guilty plea," Reuters, March 11, 2009, http://www .reuters.com/article/2009/03/11/us-madoff-idUSTRE 52A5JK20090311?pageNumber=2&virtualBrand Channel=0&sp=true (accessed February 14, 2012). **Enron, the scandal-plagued Texas corporation:** Roberts, Joel, "Enron Traders Caught on Tape," *CBSNEWS.COM*, December 5, 2007, http://www.cbsnews.com/8301-18563 _162-620626.html?tag=contentMain;contentBody (accessed February 10, 2012) **Enron held rights to a vital electricity transmission line:** Leopold, Jason, "Enron Linked to California blackouts," *Market Watch*, May 16, 2002, http://www.marketwatch.com/Story/story/print ?guid=4061B1B0-7DC7-4A4F-AE4A-3C119D69A93A (accessed February 10, 2012).

264 **But Enron made millions:** Roberts, Joel, "Enron Traders Caught on Tape," December 5, 2007; **"Just cut 'em off":** Enron and Ken Lay contributed heavily to George W. Bush's two campaigns for governor and first campaign for president. Even after the California energy crisis, then-president George W. Bush vetoed measures to cap energy prices in California.

16: AGI 2.0

267 **Authorizing a machine to make lethal combat decisions:** Finn, Peter, "A Future for Drones: Automated Killing," *Washington Post*, September 19, 2011, http://www .washingtonpost.com/national/national-security/a-future -for-drones-automated-killing/2011/09/15/gIQAVy9mgK

_print.html (accessed February 10, 2012). **a scary list of weaponized robots:** Arkin, Ronald, Mobile Robot Laboratory College of Computing Georgia Institute of Technology, "Governing Lethal Behavior: Embedding Ethics in a Hybrid Deliberative/Reactive Robot Architecture*," last modified 2011 (accessed February 10, 2012).

Index

Aboujaoude, Elias, 146

accidents, 27, 28, 93, 243
  AI and, *see* risks of artificial
    intelligence
  nuclear power plant, 27, 28,
    93, 155, 188–89

Adaptive AI, 39, 178, 179

affinity analysis, 73, 203

agent-based financial
    modeling, 124–26

"Age of Robots, The"
    (Moravec), 205–6

*Age of Spiritual Machines, The:
  When Computers Exceed
  Human Intelligence*
  (Kurzweil), 132, 134,
  150

AGI, *see* artificial general
    intelligence

AI, *see* artificial intelligence

AI-Box Experiment, 33–34,
    36–37, 48, 51, 64–65,
    67–68

airplane disasters, 28

Alexander, Hugh, 106

Alexander, Keith, 244

Allen, Paul, 149, 207

Allen, Robbie, 226

Allen, Woody, 161

AM (Automatic
    Mathematician),
    174

Amazon, 23, 74, 91, 116, 177,
    203, 246, 247

Anissimov, Michael, 188

anthropomorphism, 17–20,
    62, 78

apoptotic systems, 238–41

Apple, 194, 264
  iPad, 140

Apple (*continued*)
  iPhone, 142, 146, 179, 183,
    192
  Siri, 122, 179–80, 182, 183,
    193, 211, 219, 235
Arecibo message, 89–91
Aristotle, 190
artificial general intelligence
    (AGI; human-level AI):
  body needed for, 24–25, 88,
    166, 167, 173–74, 227–28,
    240–41
  definition of, xvii, 8, 17, 22,
    25
  emerging from financial
    markets, 124–29
  first-mover advantage in,
    11, 158
  jump to ASI from, 25, 31,
    32, 196, 242; *see also*
    intelligence explosion
  by mind-uploading, 62
  by reverse engineering
    human brain, 42–43, 45,
    57–58, 62, 143–45, 163, 165,
    202, 206–9, 212–18, 235
  time and funds required to
    develop, 25, 32, 37–39,
    147, 163, 169, 185–86,
    196–200, 229, 265–66
  Turing test for, 24, 65–66,
    102, 145, 205, 233, 234
artificial intelligence (AI):
  black box tools in, 75, 86,
    113–14, 128, 215–16, 230

definition of, 7
  drives in, *see* drives
  as dual use technology, 155
  emotional qualities in,
    18–19, 227–28, 266
  as entertainment, 26
  examples of, 17, 22–23
  explosive, *see* intelligence
    explosion
  friendly, *see* Friendly AI
  funding for jump to AGI
    from, 163, 185–86, 229
  Joy on, 134
  risks of, *see* risks of artificial
    intelligence
  Singularity and, *see*
    Singularity
  tight coupling in, 94, 253,
    255
  utility function of, 55, 56,
    61, 63
  virtual environments for,
    83, 129, 166, 173, 240–41
artificial neural networks
    (ANNs), xiv, 86, 112–14,
    125, 128, 166, 204,
    214–16, 230–32
artificial superintelligence
    (ASI), 8–18
  anthropomorphizing, 17–20
  gradualist view of dealing
    with, 31
  jump from AGI to, 25, 31,
    32, 196, 242; *see also*
    intelligence explosion

morality of, 12, 18
nanotechnology and, 29
runaway, 14–15, 234, 235
Artilect War, The (de Garis),
    85–86
ASI, *see* artificial
    superintelligence
Asilomar Guidelines, 236–37
ASIMO, 88
Asimov, Isaac:
    Three Laws of Robotics of,
        4–5, 20–21, 55, 58, 134,
        159
    Zeroth Law of, 21
Association for the
    Advancement of Artificial
    Intelligence (AAAI),
    159–60, 237
asteroids, 27, 58, 64
Atkins, Brian and Sabine, 36
Automated Insights, 226
availability bias, 27, 53–54

Banks, David L., 112
Bayes, Thomas, 103
Bayesian statistics, 103–4, 108
Biden, Joe, 262
biotechnology, 134, 155
black box systems, 75, 86,
    113–14, 128, 215–16, 230
Blue Brain project, 62
Bok globules, 91
Borg, Scott, 250
Bostrom, Nick, 19, 22, 56,
    160, 203, 211

botnets, 245–47
Bowden, B. V., 115
brain, 144, 146, 152, 201–2,
    207, 209
    augmentation of, *see*
        intelligence augmentation
    basal ganglia in, 217–18
    cerebral cortex in, 217–18
    neurons in, 213–15, 226
    reverse engineering of,
        42–43, 45, 57–58, 62,
        143–45, 163, 165, 202,
        206–9, 212–18, 235
    synapses in, 213, 215,
        222
    uploading into computer, 62
Brautigan, Richard, 252
Brazil, 255
Brooks, Rodney, 2, 171–72
Busy Child scenario, 7–8, 24,
    31, 33, 70, 83, 84, 100,
    123, 162, 169, 170, 176,
    233, 237, 242, 261
Butler, Samuel, 161

CALO (Cognitive Assistant
    that Learns and
    Organizes), 182–83
Carr, Nicholas, 146, 192–93
cave diving, 242
Center for Applied Rationality
    (CFAR), 59–60
Chandrashekar, Ashok,
    216
chatbots, 66, 149

chess-playing computers, 68, 70, 71, 76–77, 81, 38, 87, 96, 98, 109, 204, 206
   Deep Blue, 13, 23, 43–44, 219, 224
China, 32, 59, 200–201, 249–50
Chinese Room Argument, 45–46, 66, 222–23
Cho, Seung-Hui, 103
Church, Alonso, 107–8
Churchill, Winston, 107, 109
Church-Turing hypothesis, 107–8
Clarke, Arthur C., 1–2
climate change, 32–33, 142–43
cloud computing, 91, 116–17, 139, 177, 192, 234, 246–47
cognitive architectures, 17, 46, 66, 75, 76, 80, 166, 188, 202, 230
   OpenCog, 164–66, 173, 178, 198, 211, 240
cognitive bias, 26–27, 80, 235
Cognitive Computing, 180–81
Coherent Extrapolated Volition (CEV), 56–57
Colossus, 109
"Coming Technological Singularity, The" (Vinge), 118–19
computational neuroscience, 212–14, 222, 226, 230

computers, computing, 139
   cloud, 91, 116–17, 139, 177, 192, 234, 246–47
   detrimental effects from, 146–47
   exponential growth in power of, 130–31, 139–42
   see also programming; software
computer science, 40, 72, 164–66, 180, 212, 214, 229–30
consciousness, 45, 46
creativity, 92, 95–97, 169, 233
cybercrime, 244–64
Cyc, 166–67, 174, 176
Cycle Computing, 177, 188
Cycorp, 17, 166

DARPA (Defense Advanced Research Projects Agency), 17, 40, 57, 59, 60, 127, 154, 157, 159–60, 167, 180–83, 189, 194, 235, 238, 266
Darwin Machine, 86
Deep Blue, 13, 23, 43–44, 219, 224
de Garis, Hugo, 84–86
Dennett, Daniel, 164
Dijkstra, Edger, 223
DNA-related research, 236–37

Dongarra, Jack, 140
Drake, Francis, 89
drives, 8, 12, 18, 48, 70,
    78–79, 81–97, 156, 163
  creativity, 92, 95–97, 169,
    233
  efficiency, 82–83, 92–93,
    95, 96, 169, 176, 233
  resource acquisition, 82,
    86–93, 95, 96, 169, 233
  self-preservation, 82–86,
    92–93, 95, 96, 169, 233
Dugan, Regina, 236
Duqu, 256, 258, 259
Dyson, George, 123, 210,
    265

ecophagy, 15
efficiency, 82–83, 92–93, 95,
    96, 169, 176, 233
Einstein, Albert, 190
emotions, 18–19, 62, 227–28,
    266
energy grid, 248, 252–56,
    260, 262, 264
Enigma, 107–9
Enron, 263–64
Eurisko, 174–76
evil, 157
extropians, 149

Fastow, Andrew, 263
Ferrucci, David, 149, 223,
    232–33
financial scandals, 263

financial system, 94–95,
    124–29, 184, 253
Flame, 256, 258, 259
Foreign Affairs, 251
Freidenfelds, Jason, 40–41
Friendly AI, 18, 46, 48,
    51–64, 66, 68, 75, 100,
    153–54, 169, 199–201,
    230, 241
  Coherent Extrapolated
    Volition and, 56–57
  definition of, 55
  intelligence explosion and,
    60–61
  SyNAPSE and, 57–58
Future of Humanity Institute,
    158, 203, 211, 243

genetic algorithms, 74–75,
    113, 174, 188, 215,
    230–32
genetic engineering, 138, 150,
    155
genetic programming, 74–75,
    82, 128, 166
George, Dileep, 114
global warming, 32–33,
    142–43
Global Workspace Theory,
    198
Goertzel, Benjamin, 30, 31,
    120, 164–66, 167–73,
    178, 183–85, 196–200,
    202, 236, 240–41
Golden Rule, 20

Good, I. J., xx, 17, 101–6,
        108–17, 118–19, 121–22,
        129, 131, 133, 138, 162,
        163, 172, 176, 196
Google, 17, 39–42, 59, 91, 116,
        122, 127, 152, 163, 177,
        182, 194–96, 210, 236,
        246, 249, 266
    search engine, 23, 99,
        189–93, 195
Google X, 41–42
Granger, Richard, 43, 207–9,
        212–13, 216–18, 220,
        222, 227
Grassie, William, 137–38
Greaves, Mark, 207
Grossberg, Steven, 114
grounding problem, 174
GUI (Graphical User
        Interface), 180, 194

hackers, 244–64
Hawking, Stephen, 148,
        241
Hawkins, Jeff, 114, 178
Hebb, Donald, 112–13
heuristics, 27
Hibbard, Bill, 187
high-frequency trading systems
        (HFTs), 94–95, 127
Hillis, Danny, 69
Holtzman, Golde, 103,
        111–12
Horvitz, Eric, 159
Hughes, James, 63–64

I, Robot (Asimov), 20
IA, see intelligence
        augmentation
IBM, 40, 59, 113, 114, 127,
        163, 216, 236
    Blue Brain, 62
    Deep Blue, 13, 23, 43–44,
        219, 224
    SyNAPSE, 57–58, 80, 182,
        216, 235
    Watson, xix, 13–14, 17, 23,
        122, 149, 155, 187, 204,
        211, 219–24, 232–33,
        266
immortality, 122, 131, 133,
        137, 143, 145–46, 149,
        152, 160
incomprehensibility, 95
Ings, Simon, 210
inscrutability paradox,
        230–31
Institute for Ethics and
        Emerging Technologies
        (IEET), 63, 158
integrated circuits, 139–42
intelligence, 23–24, 124, 202,
        212–13
    embodiment and, 24, 88,
        227–28, 240–41
    emerging from Internet,
        122–24
    knowledge and, 190
intelligence augmentation
        (IA), 123, 145–47, 152,
        156–58, 189–95

intelligence explosion, 8, 17,
      29, 32, 51, 60–61, 68,
      99–101, 104–5, 111–17,
      129, 152, 158, 159,
      162–64, 168, 169, 172–73,
      185, 189, 195, 266
   economics and, 183–86,
      235–36
   financial markets and,
      128–29
   hard takeoff in, 31, 32,
      169–70
   limiting factors to, 177–78,
      185, 229
   Moore's law and, 140–41
   risks of, *see* risks of artificial
      intelligence
   self-awareness and, *see*
      self-awareness
   self-improvement and, *see*
      self-improvement
   software complexity and,
      163, 186, 189, 196, 199,
      201, 202, 229
   system space for, 177
Internet, 14, 30, 139, 140, 167,
      173, 180, 190–92
   intelligence emerging from,
      122–24
   security and, 244–64
iPad, 140
iPhone, 142, 146, 179, 183,
      192
   Siri, 122, 179–80, 182, 183,
      193, 211, 219, 235

Iran, 38, 200, 257–59, 261
iRobot, 171, 267

Jennings, Ken, 223–24
Jeopardy!, 13, 23, 204, 205,
      219–24
Joy, Bill, 134, 151, 160
Jurvetson, Steve, 230–31

Kahneman, Daniel, 26–27
Kasparov, Gary, 43–44
Kelly, Kevin, 123–24
Koza, John, 74–75
Kroft, Steve, 260
Kurzweil, Ray, xx, 2, 28–31,
      116, 130–31, 144–47,
      149–56, 160, 162–63,
      165, 171, 196, 197, 200,
      207, 218–19, 222, 236,
      242, 243, 244–45, 252,
      265
   global warming problem as
      viewed by, 142–43
   Google and, 42
   Law of Accelerating
      Returns of, 131, 138–39,
      142–43, 147, 152, 162,
      176–77, 246
   Singularity and, 28–30,
      46–47, 119, 122, 132–35,
      138, 145, 147, 149, 150,
      152, 160, 162

Langner, Ralph, 260–61
Lanier, Jaron, 147

Law of Accelerating Returns
    (LOAR), 131, 138–39,
    142–43, 147, 152, 162,
    176–77, 246
Lay, Kenneth, 263
Leibnitz, Gottfried, 197–98
Lenat, Douglas, 167, 174–76
Lewis, H. W., 32, 64
LIDA (Learning Intelligent
    Distributed Agent),
    178–79, 183, 198, 211
Lifeboat Foundation, 243
Lipson, Hod, 174, 204
Loebner, Hugh, 66
Lynn, William J., III, 154,
    247–51, 254–55

Machine Intelligence
    Research Institute
    (MIRI), xxi, 22, 35, 36,
    47, 49, 50, 59–60, 64, 92,
    99, 158, 169, 188, 243
    Singularity Summit, 36,
    148–49
machine learning, 73–74,
    112–14, 204
Madoff, Bernie, 263
malware, 244–64
Mazzafro, Joe, 248
McCarthy, John, 204
McGurk, Sean, 260
military, 21, 25, 60, 171–72, 184
    battlefield robots and
        drones, 21, 60, 154, 171,
        235, 266–67

DARPA, see DARPA
    energy infrastructure and,
        254
    nuclear weapons, see
        nuclear weapons
Mind Children (Moravec), 204
Minsky, Marvin, 66
Mitchell, Tom, 159–60
mobile phones, 192, 194–95
    see also iPhone
Monster Cat, 177, 188
Moore, Gordon, 139, 141–42
Moore's Law, 139–42, 146,
    147
morality, 12, 18, 61, 66,
    157–58, 170–71
    see also Friendly AI
Moravec, Hans, 204
Moravec's Paradox, 204
mortality, see immortality
mortgage crisis, 263
Mutually Assured Destruction
    (MAD), 123, 154

nano assemblers, 15–16
nanotechnology, 15, 28–29,
    32, 37, 49, 63, 83, 85,
    88–89, 132, 134, 138,
    146, 150, 155, 170, 200
    "gray goo" problem and, 29
natural language processing
    (NLP), 167, 173, 178–80,
    219, 224
natural selection, 54, 74, 124,
    199

Nekomata (Monster Cat), 177, 188

NELL (Never-Ending-Language-Learning system), 167, 182

neural networks, 86, 112–14, 125, 128, 166, 204, 214–16, 230–32

neurons, 213–15, 226

New Scientist, 198

New York Times, 41, 258, 261, 262

Newman, Max, 109

Newton, Isaac, 162, 174, 197–98

Ng, Andrew, 42

9/11 attacks, 27–28, 38, 243, 263

Normal Accidents: Living with High-Risk Technologies (Perrow), 93–94, 188, 229

normalcy bias, 188

North Korea, 38, 200

Norvig, Peter, 39–40, 196

Novamente, 17, 164

nuclear fission, 21

nuclear power, 155, 262
    plant disasters, 27, 28, 93, 155, 188–89

nuclear weapons, 27, 33, 38, 84, 154–55, 162, 163, 236, 250
    of Iran, 258–59, 261

Numenta, 17, 114, 178

Ohana, Steve, 95

Olympic Games (cyberwar campaign), 258–59

Omohundro, Stephen, 68, 70–73, 75–77, 79–83, 86–87, 91, 95, 97, 98, 101, 115, 153, 156, 158, 160, 162, 163, 169, 172, 184, 200, 201, 233, 234, 240

OpenCog, 164–66, 173, 178, 198, 211, 240

Otellini, Paul, 132

Page, Larry, 40, 41, 142–43, 190, 193

paper clip maximizer scenario, 56

parallel processing, 213, 214, 216, 220

pattern recognition, 114, 165, 215

Pendleton, Leslie, 110–11, 116, 119

Perceptron, 113, 114

Perrow, Charles, 93–95, 188, 229, 253

Piaget, Jean, 166

power grid, 248, 252–56, 260, 262, 264

Precautionary Principle, 151

programming, 71, 81, 153, 184
    bad, 72
    evolutionary, 75, 86

programming (*continued*)
  genetic, 74–75, 82, 128, 166
  ordinary, 74, 230
  self-improving, *see* self-
    improvement

Rackspace, 91, 116–17, 177,
    246
rational agent theory of
    economics, 79–80
recombinant DNA, 236–37
*Reflections on Artificial
    Intelligence* (Whitby),
    231–32
resource acquisition, 82,
    86–93, 95, 96, 169, 233
risks of artificial intelligence,
    32–33, 63–64, 151–60,
    172, 200–201, 229–43
  apoptotic systems and,
    238–41
  Asilomar Guidelines and,
    236–37
  Busy Child scenario and, *see*
    Busy Child scenario
  defenses against, 158,
    241–42
  lack of dialogue about,
    25–30, 267
  malicious AI, 244–64
  Precautionary Principle
    and, 151
  runaway AI, 14–15, 234, 235
  Safe-AI Scaffolding
    Approach and, 240

  Stuxnet and, 256–62, 266,
    267
  unintended consequences,
    55–56, 96–98
robots, robotics, 23, 25, 138,
    150, 170
  Asimov's Three Laws of,
    4–5, 20–21, 55, 58, 134,
    159
  in dangerous and service
    jobs, 184–85
  in sportswriting, 225–26
Rosenblatt, Frank, 113,
    114
Rowling, J. K., 133
Rubin, Andrew, 194–95
"Runaround" (Asimov), 20

Safe-AI Scaffolding Approach,
    240
Sagan, Carl, 89
SCADA (supervisory control
    and data acquisition)
    systems, 256–59
Schmidt, Eric, 193
Schwartz, Evan, 192
Scientist Speculates, The (Good,
    ed.), 115–16
Searle, John, 45–46, 222–23
self-awareness, 4, 8, 12, 68,
    71–76, 78, 79, 81, 83,
    87–88, 100, 156, 163, 169,
    172, 176, 195, 200, 201,
    233, 241
Self-Aware Systems, 17, 68

self-improvement, 68, 72–76, 78–82, 87–88, 99, 100, 156, 163, 169, 172–73, 176, 195, 200, 201, 233, 241

self-preservation, 82–86, 92–93, 95, 96, 169, 233

September 11 attacks, 27–28, 38, 243, 263

serial processing, 213

SETI (Search for Extraterrestrial Intelligence), 89–91

Shostak, Seth, 90–92

Silicon Valley, 49–50

Singularitarians, 122, 135–38, 158

Singularity, 28–29, 60, 104–5, 118–23, 130–31, 133, 135–38, 141, 159, 173
    definitions of, 28, 46, 104, 119, 130, 133, 169
    Kurzweil and, 28–30, 46–47, 119, 122, 132–35, 138, 145, 147, 149, 150, 152, 160, 162
    technological, 46–47, 118–22

Singularity Is Near, The (Kurzweil), 29–30, 132, 144, 150, 160, 242

Singularity Summit, 36, 148–49

Singularity University, 136

Sir Groovy, 37

Siri, 122, 179–80, 182, 183, 193, 211, 219, 235

60 Minutes, 260

Skilling, Jeffrey, 263

Smart Action, 178

smart phones, 192, 194–95
    see also iPhone

software, 184
    complexity of, 163, 186, 189, 196, 199, 201, 202, 229
    malware, 244–64
    see also programming

solar energy, 142–43

space exploration, 87–92

"Speculations Concerning the First Ultraintelligent Machine" (Good), 104–5, 112, 113, 115, 117

speech recognition, 218–19

SRI International, 182, 183

stealth companies, 39, 41–42, 59, 151, 178–80, 185

Sterrit, Roy, 239–40

Stibel, Jeff, 30

Stuxnet, 256–62, 266, 267

subprime mortgage crisis, 263

Symantec, 245, 249–50

SyNAPSE, 57–58, 80, 182, 216, 235

Technological Risk (Lewis), 32

technology journalism, 26

Terminator movies, 17, 26

terrorism, 38, 58, 260, 263
    9/11 attacks, 27–28, 38, 243, 263

Thiel, Peter, 39, 149
Thinking Machines, Inc., 69
Three Mile Island, 27, 93, 188
tightly coupled systems, 94,
    253, 255
Thrun, Sebastian, 42
transhumans, 149, 156
transistors, 139–42
Traveller Trillion Credit
    Squadron, 175
Turing, Alan, 24, 65–66, 102,
    107–12, 117
Turing machine, 107–8
Turing test, 24, 65–66, 102,
    145, 205, 233, 234
Tversky, Amos, 26
two-minute problem, 29
2001: A Space Odyssey, 17

Ulam, Stanislaw, 119
utility function, 55, 56, 61, 63,
    79–80

Vassar, Michael, 35–39,
    42–43, 47–48, 59
Vicarious Systems, 17, 114
Vinge, Vernor, 118–23,
    129–31, 133, 135, 138,
    148, 189, 197, 200, 201
violence, 157, 171
Virginia Tech Massacre, 103
Virtually You (Aboujaoude), 146
voice recognition, 218–19

von Neumann, John, 79–80,
    119, 123
Voss, Peter, 39, 178, 179, 190

Wallach, Wendall, 229, 243
Wall Street, 94–95, 124–29
Warwick, Kevin, 78
Washington Post, 154
Watson, 13–14, 17, 23, 122,
    149, 155, 187, 204, 211,
    219–24, 232–33, 266
weapons, see military
Whitby, Blay, 231–32
"Why the Future Doesn't
    Need Us" (Joy), 134
Wired for Thought (Stibel), 30
Wissner-Gross, Alexander D.,
    126–29
Wolfram, Stephen, 149
Wozniak, Steve, 205

You Are Not a Gadget: A
    Manifesto (Lanier), 147
Yudkowsky, Eliezer, xv, 22,
    36–37, 48, 49–56, 60–65,
    67–68, 99, 124, 141, 149,
    153, 160, 176, 199, 200
Yudkowsky, Yehuda, 50

Zeitgist '06, 40
zero day vulnerabilities, 257
Zeroth Law, 21
Zeus malware, 244, 245, 247